Atlas
of
Embryonic Development

Atlas
of
Embryonic Development

Steven B. Oppenheimer
Richard L. C. Chao
California State University, Northridge

Allyn and Bacon, Inc.
Boston   London   Sydney   Toronto

Library of Congress Cataloging in Publication Data

Oppenheimer, Steven B., 1944-
    Atlas of embryonic development.

    1. Embryology—Atlases. I. Chao, Richard L. C.
II. Title. [DNLM: 1. Embryology—Atlases. QS 617 062a]
QL956.O658 1984   597.6′0433   83-27152
ISBN 0-205-08099-5

Printed in the United States of America.

10 9 8 7 6 5 4 3 2 1   88 87 86 85 84

Contents

Preface

This atlas includes light micrographs and scanning electron micrographs of specimens used in most developmental biology and embryology courses. We include the scanning micrographs to show three-dimensional qualities such as texture and depth that would not be apparent from light micrographs alone. All figures are labeled for easy study, and transverse sections include diagrams that indicate the position of the section in the whole embryo. Developmental timetables and sketches included should help students understand the sequence of events occurring during embryonic development from a structural standpoint. In addition, drawings have been placed in appropriate sections of this atlas that will help students grasp the nature of the tissue movements and rearrangements that lead to the observed structural changes that are characteristic of different developmental stages. Finally, a glossary of terms is included that will facilitate the definition of structures labeled in the micrographs.

We thank Mary Beth Finch, Jim Smith, John Gilman, Vicky Prescott and Sandi Kirshner for excellent assistance in the development and production of this book and the entire staff of Allyn and Bacon for concerted efforts in the final stages of its production and distribution.

Spermatogenesis and Oogenesis

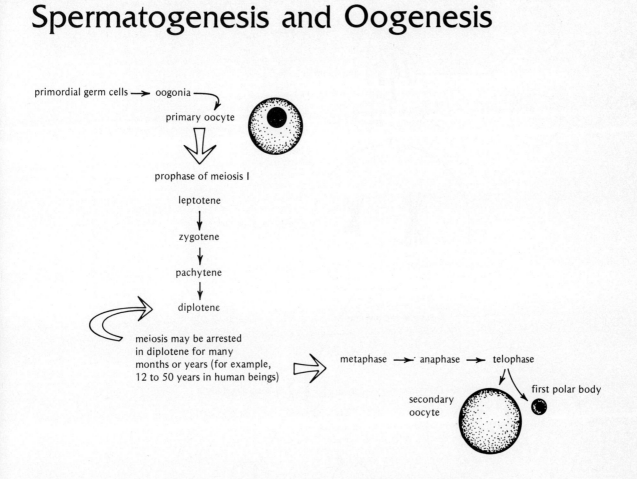

Figure 1. Meiosis I in vertebrate oogenesis. From S. B. Oppenheimer, *Introduction to Embryonic Development,* 2nd Ed. (Boston: Allyn and Bacon, 1984).

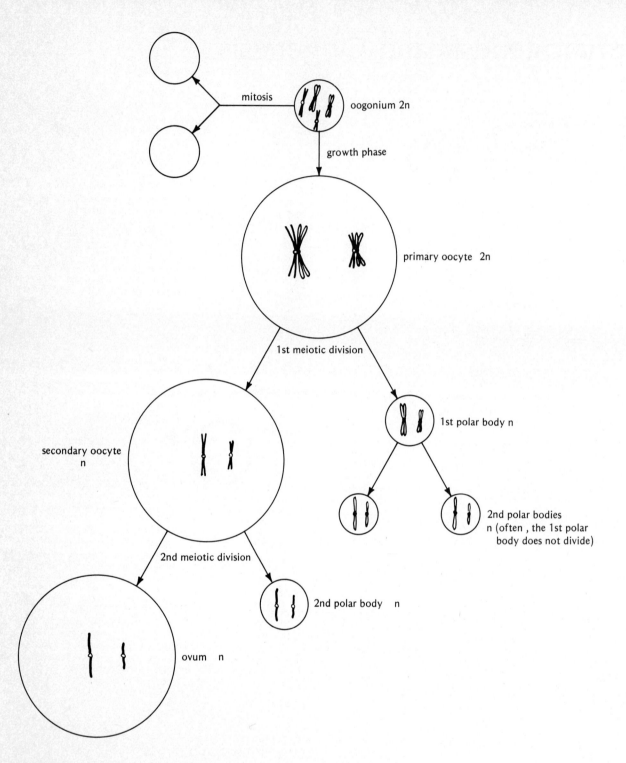

Figure 2.   Summary of oogenesis. From S. B. Oppenheimer, *Introduction to Embryonic Development,* 2nd Ed. (Boston: Allyn and Bacon, 1984).

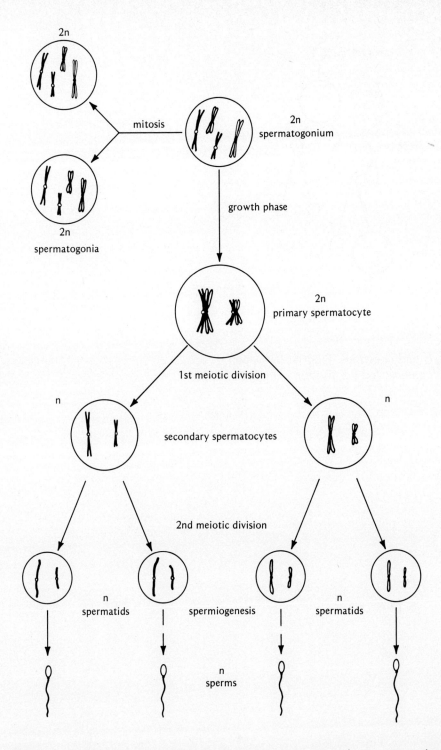

Figure 3. Summary of spermatogenesis. From S. B. Oppenheimer, *Introduction to Embryonic Development,* 2nd Ed. (Boston: Allyn and Bacon, 1984).

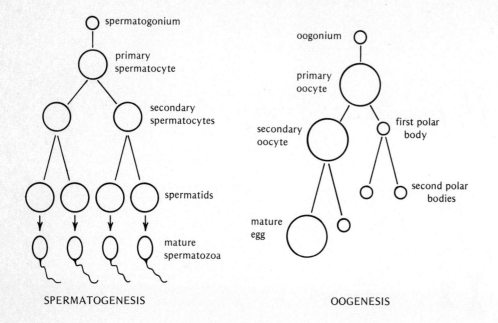

SPERMATOGENESIS   OOGENESIS

Figure 4. Comparison of spermatogenesis and oogenesis. From S. B. Oppenheimer, *Introduction to Embryonic Development,* 2nd Ed. (Boston: Allyn and Bacon, 1984).

I
The Frog

TABLE 1. Development of *Rana pipiens* at 18°C

Stage number	Age in hours	Size in millimeters	Characteristics
1	0	1.75	unfertilized egg
2	1.0	1.8	gray crescent
3	3.5	1.8	2 cells
4	4.5	1.8	4 cells
5	5.7	1.8	8 cells
6	6.5	1.8	16 cells
7	7.5	1.8	32 cells
8	16	1.8	early blastula
9	21	1.8	late blastula
10	26	1.8	early gastrula (dorsal lip)
11	34	1.9	middle gastrula (crescent blastopore)
12	42	2.0	late gastrula (yolk plug)
13	50	2.2	neural plate
14	62	2.3	neural folds
15	67	2.4	cilia rotation begins
16	72	2.5	complete neural tube
17	84	3.0	tail bud
18	96	4.0	muscular movement
19	118	5.0	heart beats
20	140	6.0	gill circulation and hatching
21	162	7.0	mouth open; cornea is transparent
22	192	8.0	tail fin circulation
23	216	9.0	opercular fold; teeth
24	240	10.0	operculum closed on right
25	284	11.0	operculum complete

[Based on results of W. Shumway, *Anat. Rec.* 78: 139–148 (1940)]

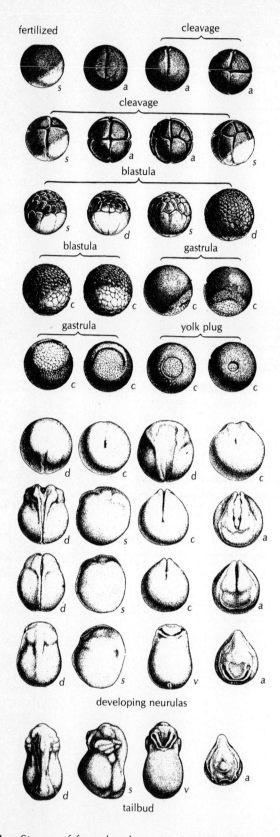

tailbud

gill buds

hatching,
gill circulation

mouth open

Table 1.   Stages of frog development *(continued)*. *a*, View from animal pole
(frontal view); *c*, caudal (blastoporal) view; *d*, dorsal view; *s*, left lateral view;
v, ventral view. (From *Development of the Vertebrates* by Emil Witschi.
Copyright 1956 by W. B. Saunders Co. Used with permission of the W. B.
Saunders Co.)

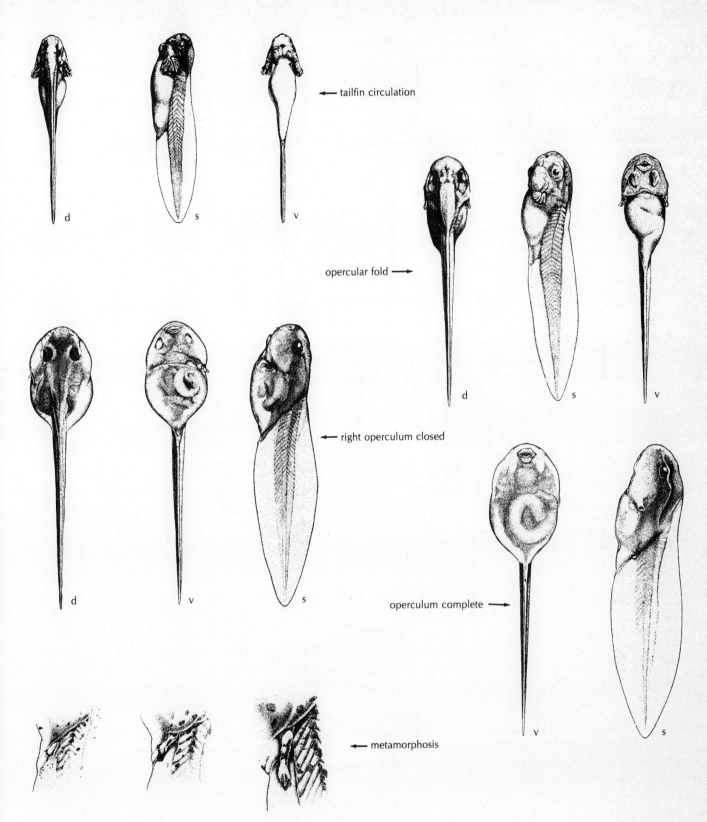

← tailfin circulation

opercular fold →

← right operculum closed

operculum complete →

← metamorphosis

Table 1.   Stages of frog development *(continued)*. *a*, View from animal pole (frontal view); *c*, caudal (blastoporal) view; *d*, dorsal view; *s*, left lateral view; *v*, ventral view. (From *Development of the Vertebrates* by Emil Witschi. Copyright 1956 by W. B. Saunders Co. Used with permission of the W. B. Saunders Co.)

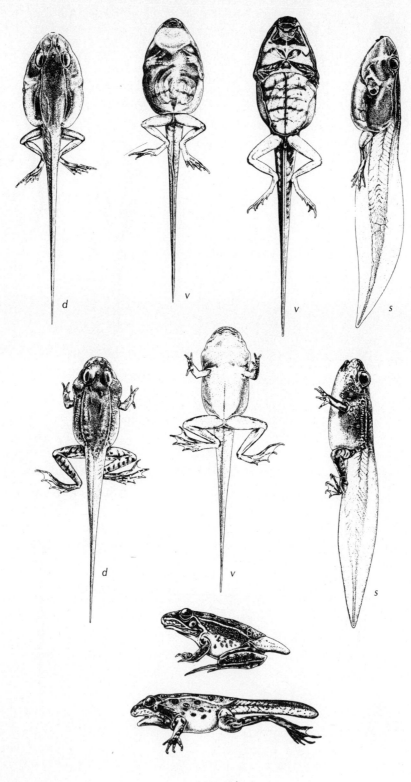

metamorphosis

Table 1.   Stages of frog development *(continued)*. *a,* View from animal pole (frontal view); *c,* caudal (blastoporal) view; *d,* dorsal view; *s,* left lateral view; *v,* ventral view. (From *Development of the Vertebrates* by Emil Witschi. Copyright 1956 by W. B. Saunders Co. Used with permission of the W. B. Saunders Co.)

A. Frog Gametes

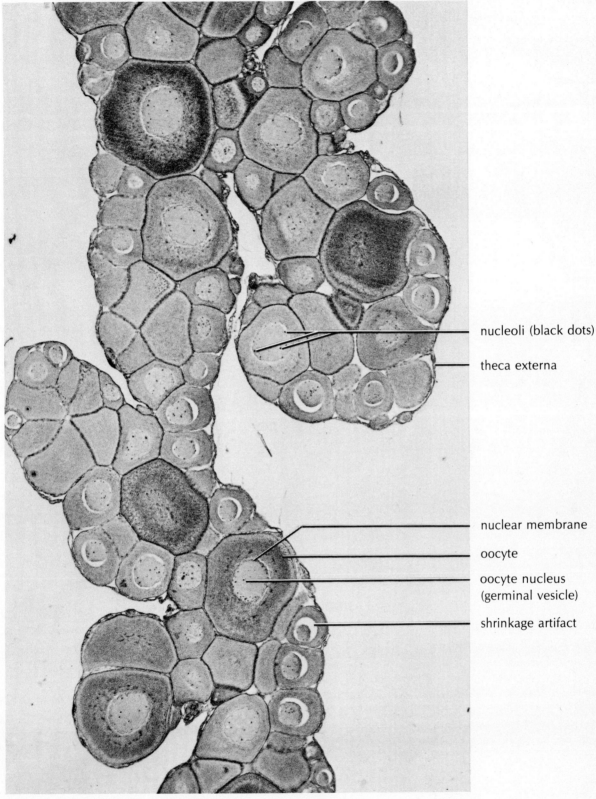

nucleoli (black dots)

theca externa

nuclear membrane

oocyte

oocyte nucleus
(germinal vesicle)

shrinkage artifact

Figure 5. Frog ovary. (66.6×)

nuclei of
follicle cells   oocyte cytoplasm   nuclear
membrane

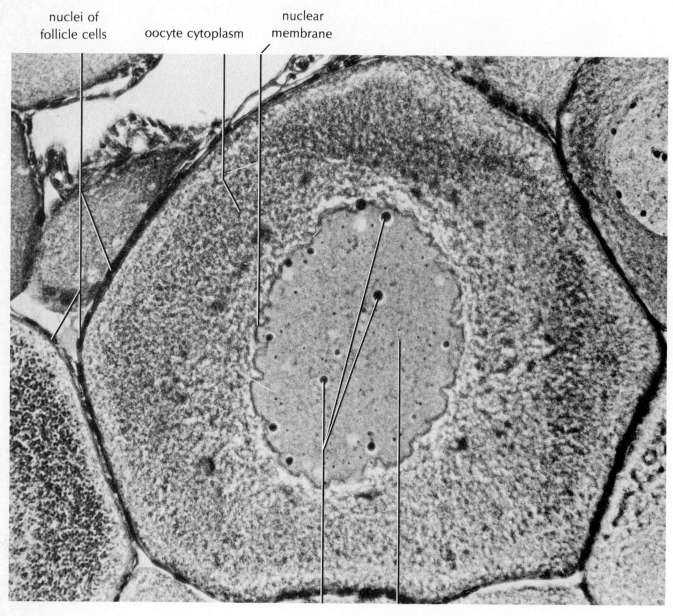

nucleoli   nucleus (germinal vesicle)

Figure 6. Frog oocyte, enlarged. (621.6×)

seminiferous tubules

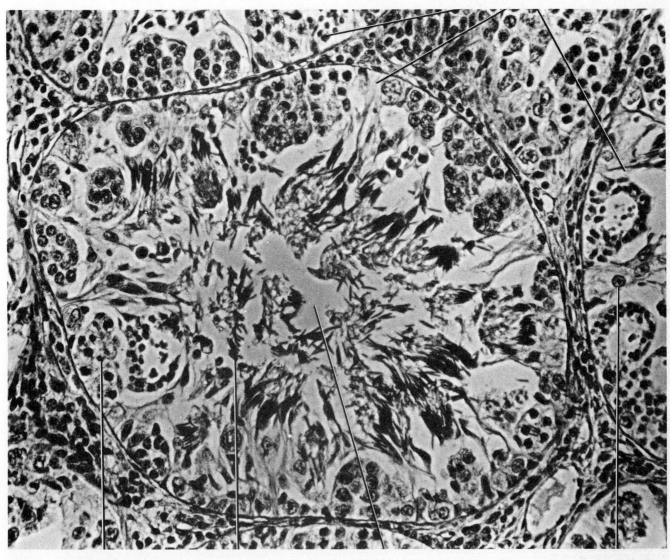

spermatocytes   sperm   lumen   Sertoli cell

Figure 7. Frog testis showing seminiferous tubule. (404.8×)

B. Frog Blastula and Gastrula

animal region   blastocoel

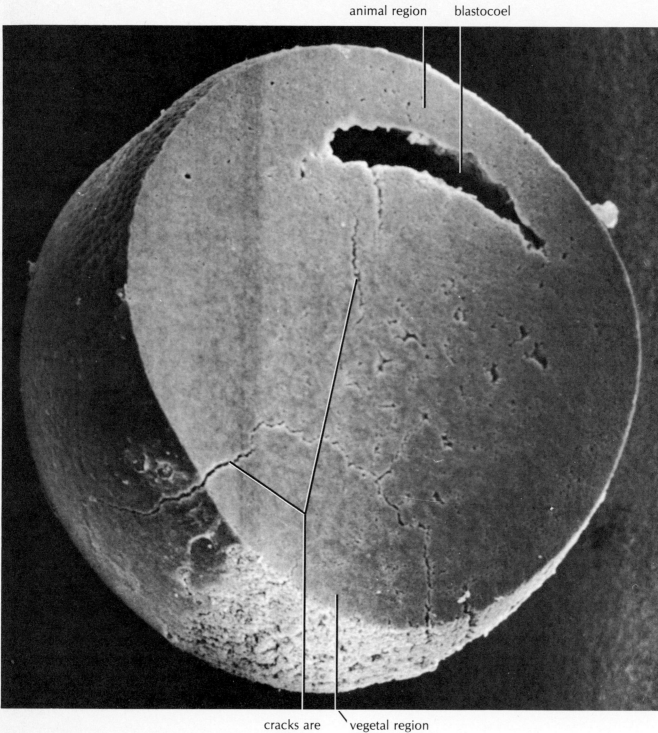

cracks are
artifacts

vegetal region

Figure 8. Early frog blastula, scanning electron micrograph. (137.2×)

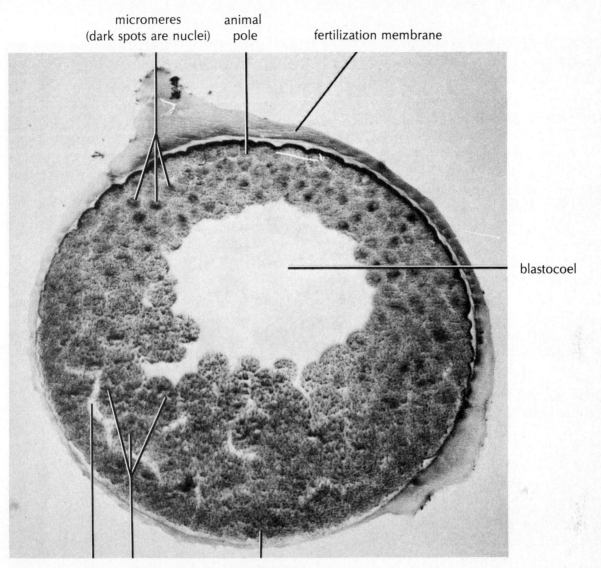

micromeres
(dark spots are nuclei)

animal
pole

fertilization membrane

blastocoel

shrinkage artifact   macromeres   vegetal pole

Figure 9.   Frog blastula. (92.5×)

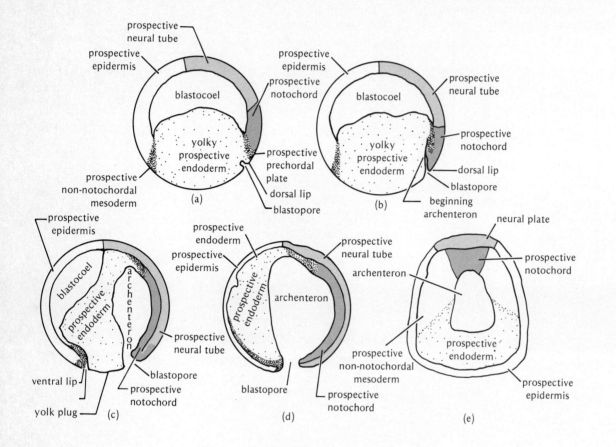

Figure 10. Gastrulation in the amphibian. Keller and colleagues, working with *Xenopus,* a frog, found that prospective mesoderm may be located in deeper layers of the blastula instead of in surface regions. After *W. Vogt., Roux Arch. 120,* 385–706 (1929). From S. B. Oppenheimer, *Introduction to Embryonic Development,* 2nd Ed. (Boston: Allyn and Bacon, 1980).

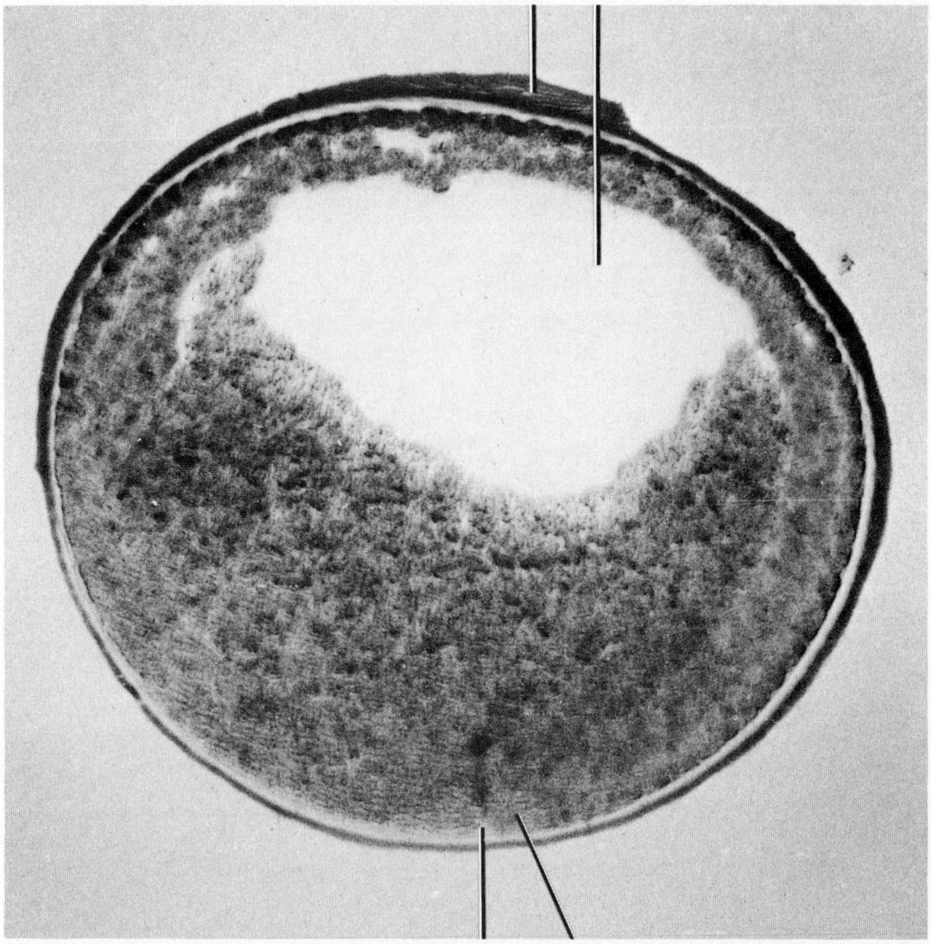

fertilization membrane   blastocoel

blastopore   dorsal lip of blastopore

Figure 11.   Early frog gastrula. (92.5×)

archenteron roof   archenteron

dorsal lip
of blastopore

yolk plug

ventral lip
of blastopore

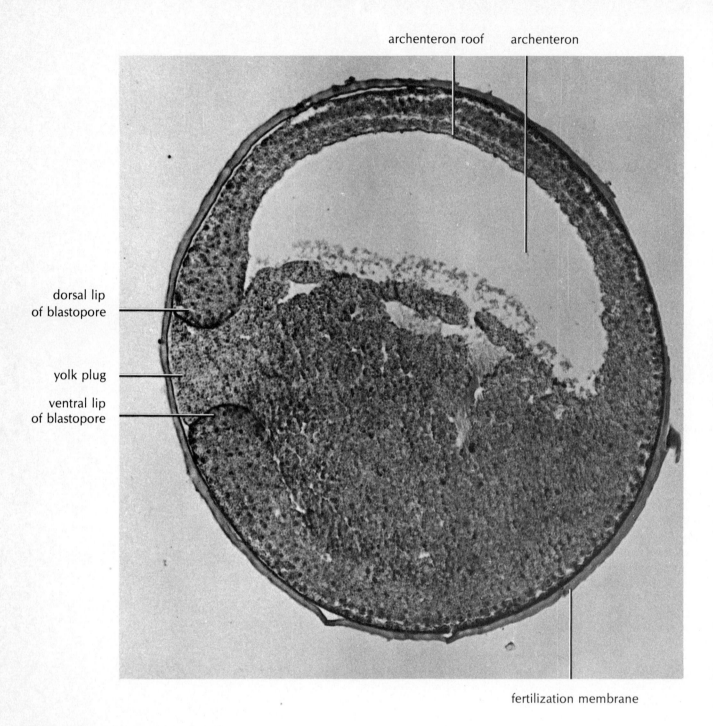

fertilization membrane

Figure 12.   Frog yolk plug. (97.5×)

C. Frog Neurula

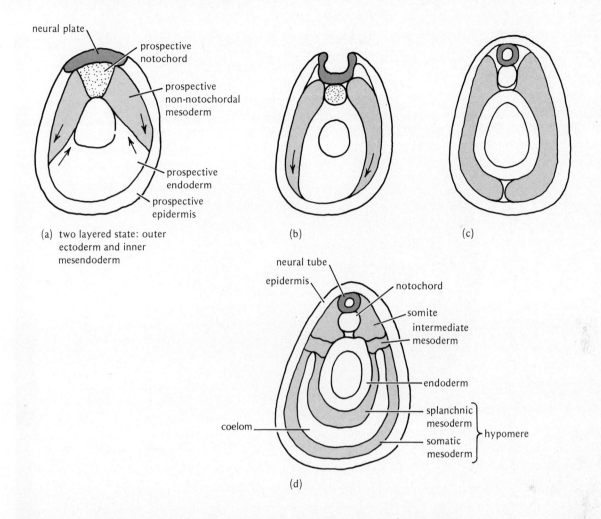

(a) two layered state: outer ectoderm and inner mesendoderm

neural plate
prospective notochord
prospective non-notochordal mesoderm
prospective endoderm
prospective epidermis

(b)

(c)

(d)

neural tube
epidermis
notochord
somite
intermediate mesoderm
endoderm
splanchnic mesoderm
somatic mesoderm
} hypomere
coelom

Figure 13. Formation of the three-layered state in the amphibian. Coelom formation occurs by mesodermal splitting. From S. B. Oppenheimer, *Introduction to Embryonic Development,* 2nd Ed. (Boston: Allyn and Bacon, 1984).

neural folds

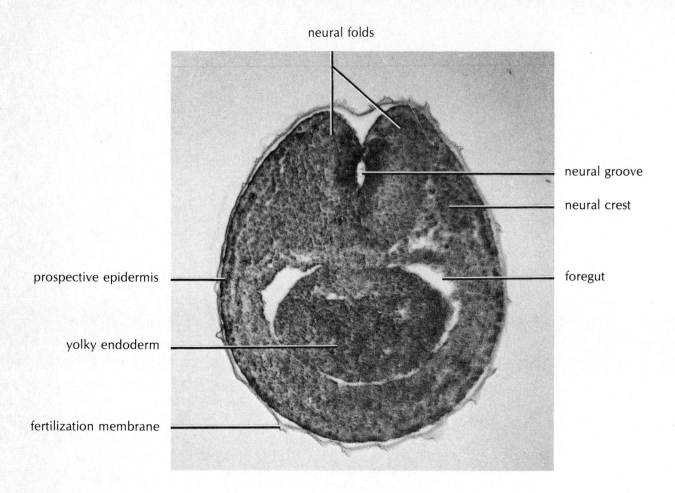

neural groove

neural crest

prospective epidermis

foregut

yolky endoderm

fertilization membrane

Figure 14.   Frog embryo, neural fold stage, transverse section through foregut. (185×)

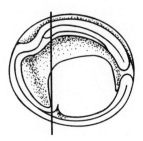

neural groove   neural fold   fertilization membrane

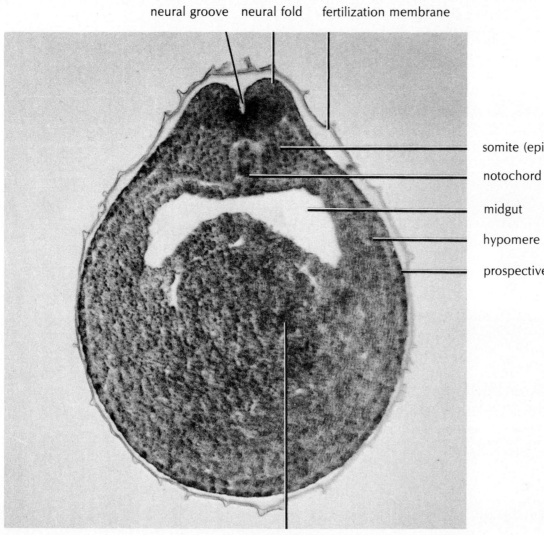

somite (epimere)

notochord

midgut

hypomere

prospective epidermis

yolky endoderm

Figure 15.   Frog embryo, neural fold stage, transverse section through midgut. (185 ×)

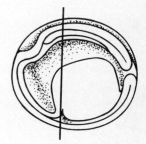

fertilization membrane   neural groove

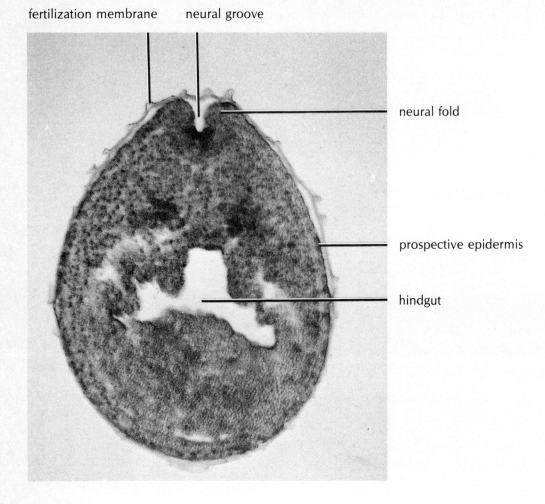

neural fold

prospective epidermis

hindgut

Figure 16.   Frog embryo, neural fold stage, transverse section through hindgut. (185×)

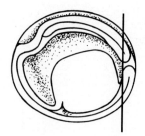

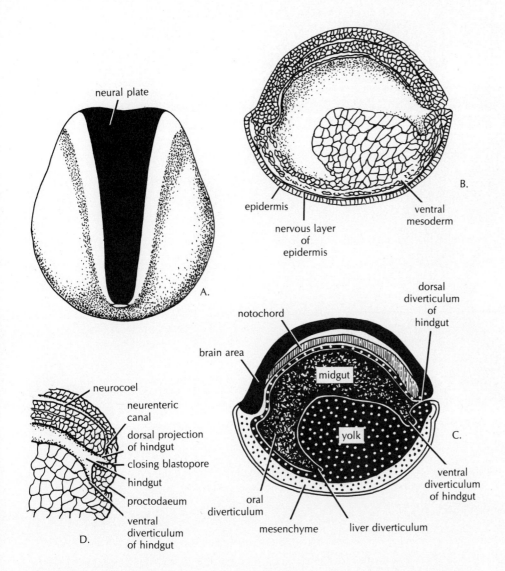

Figure 17.   Beginning neural fold stage of frog embryo from prepared material. (A) Beginning neural fold stage as seen from dorsal view. (B) Sagittal section near median plane of embryo similar to that shown in (A). (C) Same as (B), showing organ-forming areas. (D) Midsagittal section of caudal end of frog embryo. Observe that the blastopore practically is closed, while the dorsal diverticulum of the hindgut connects with the neurocoel to form the neurenteric canal. Observe, also, ventral diverticulum of hindgut. From O. E. Nelsen, *Comparative Embryology of the Vertebrates* (New York: McGraw-Hill, 1953).

brain (neural tube)   mesoderm

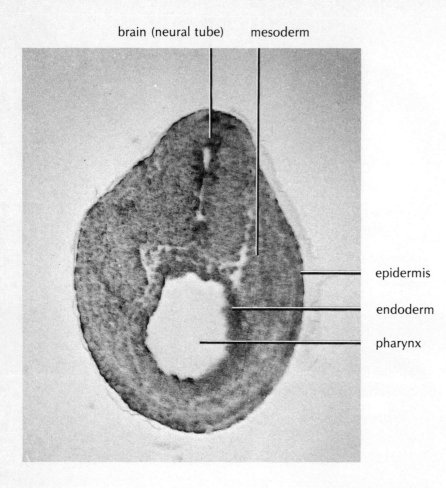

epidermis

endoderm

pharynx

Figure 18. Frog embryo, neural tube stage, transverse section through foregut. (185×)

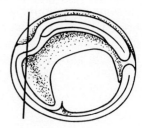

rhombencephalon (neural tube)

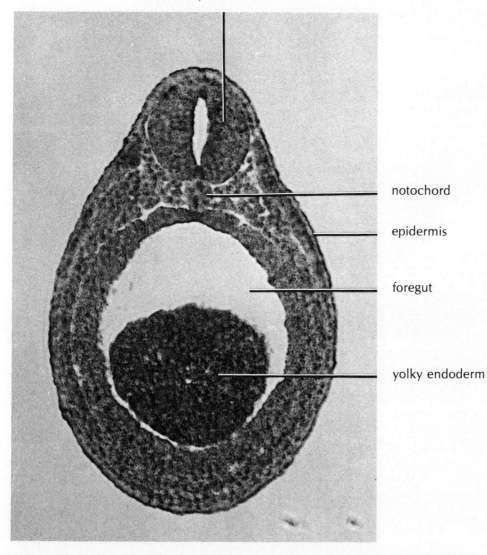

notochord

epidermis

foregut

yolky endoderm

Figure 19. Frog embryo, neural tube stage, transverse section through foregut. (106×)

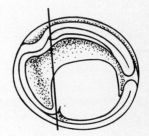

spinal cord (neural tube)

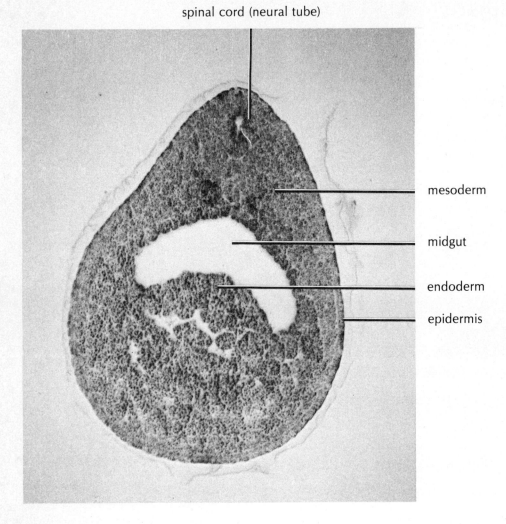

mesoderm

midgut

endoderm

epidermis

Figure 20. Frog embryo, neural tube stage, transverse section through midgut. (185 ×)

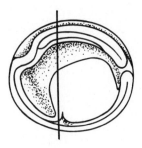

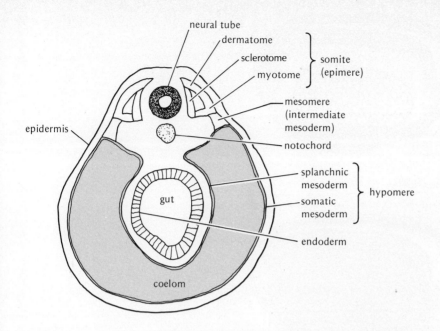

Figure 21.   Mesoderm divisions in typical vertebrate neurula. From S. B. Oppenheimer, *Introduction to Embryonic Development,* 2nd Ed (Boston: Allyn and Bacon, 1984).

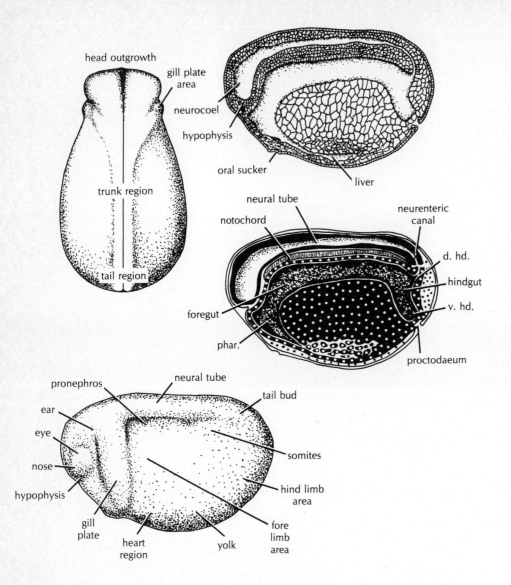

Figure 22. Early neural tube stage of the frog, *Rana pipiens,* 2½ to 3 mm. in length. (A) Dorsal view. (B) Midsagittal section of embryo similar to (A). (C) Same as (B), showing organ-forming areas. Abbreviations: V. HD. = ventral hindgut diverticulum; D. HD. = dorsal hindgut diverticulum; PHAR. = pharyngeal diverticulum of foregut. (D). Lateral view of (A). From O. E. Nelsen, *Comparative Embryology of the Vertebrates* (New York: McGraw-Hill, 1953).

D. Frog Tailbud, 4 mm Stage

notochord pharynx prosencephalon

spinal cord

notochord

tailbud

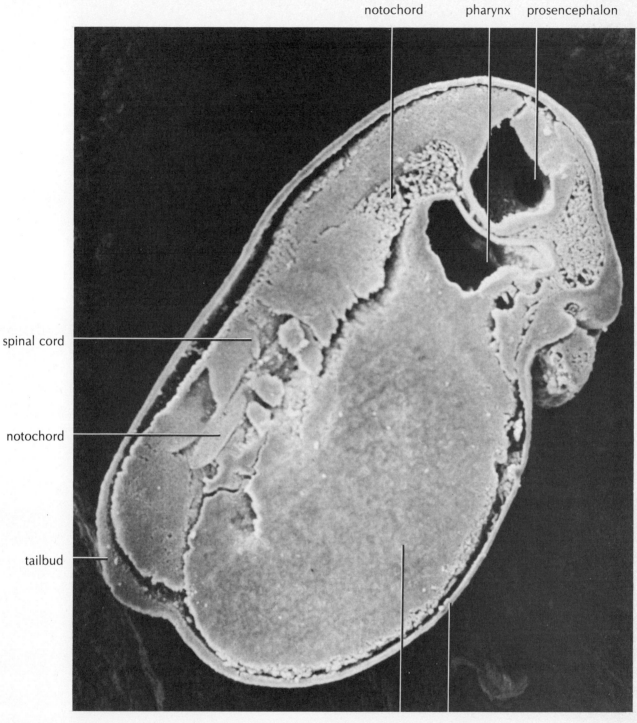

yolky endoderm epidermis

Figure 23. Frog embryo, tailbud stage, sagittal section, scanning electron micrograph. (63.9×)

tail bud   epidermis   somites   otic vesicle   optic cup

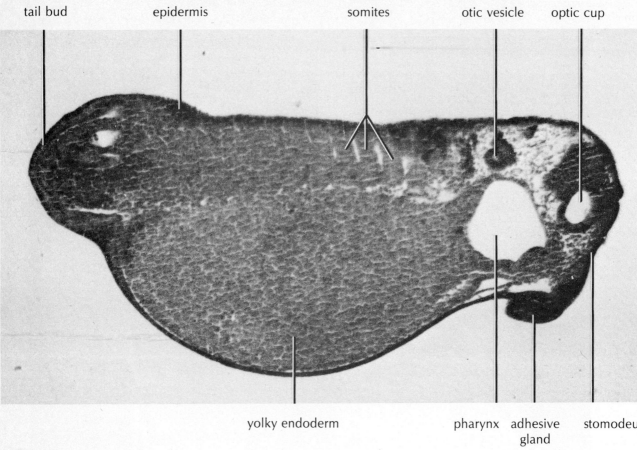

yolky endoderm   pharynx   adhesive   stomodeum
                                                        gland

Figure 24.   Frog embryo, tailbud stage, sagittal section. (44.4×)

tailbud   epidermis   somites   midgut   spinal cord   notochord   rhombencephalon

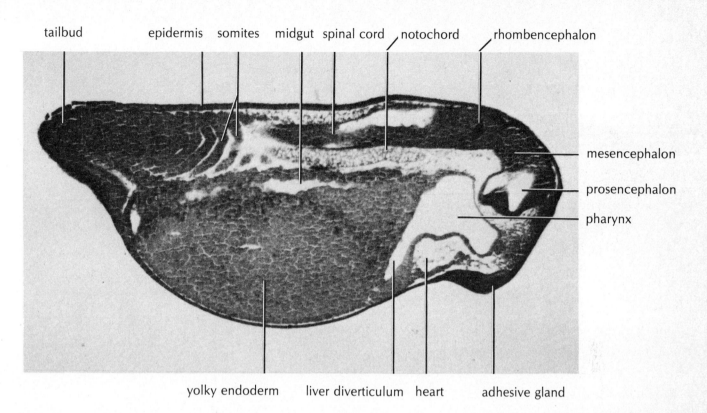

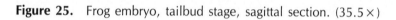

mesencephalon

prosencephalon

pharynx

yolky endoderm   liver diverticulum   heart   adhesive gland

Figure 25.   Frog embryo, tailbud stage, sagittal section. (35.5×)

epidermis   lens

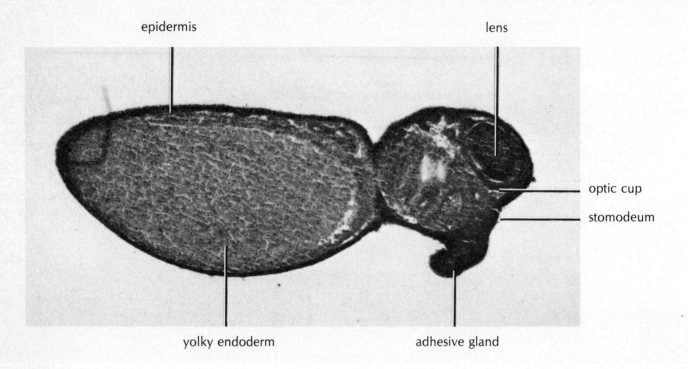

yolky endoderm   adhesive gland

optic cup

stomodeum

Figure 26. Frog embryo, tailbud stage, sagittal section near outer surface.
(44.4×)

tailbud

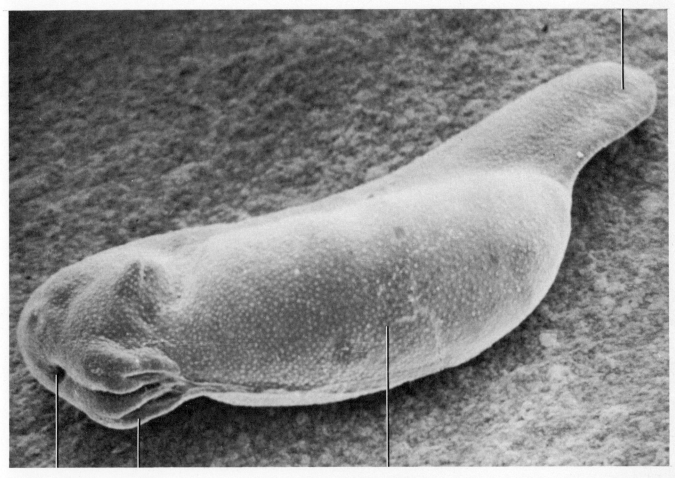

stomodeum   adhesive   yolky endoderm
               gland   (below epidermis and mesoderm)

Figure 27. Frog embryo, 4 mm, scanning electron micrograph. (67.2×)

prosencephalon   epidermis

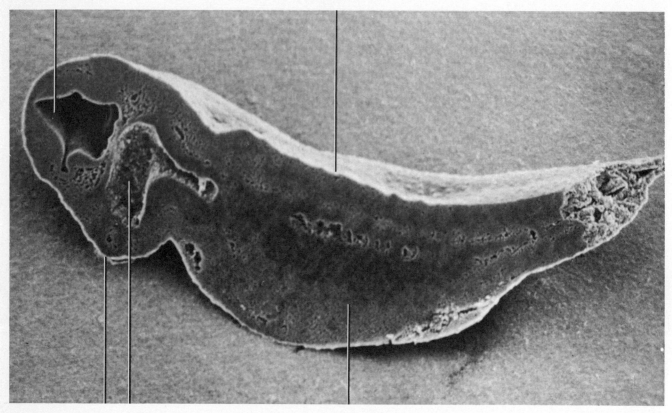

adhesive gland   pharynx   yolky endoderm

Figure 28.   Frog embryo, 4 mm, sagittal section, scanning electron micrograph. (75.2×)

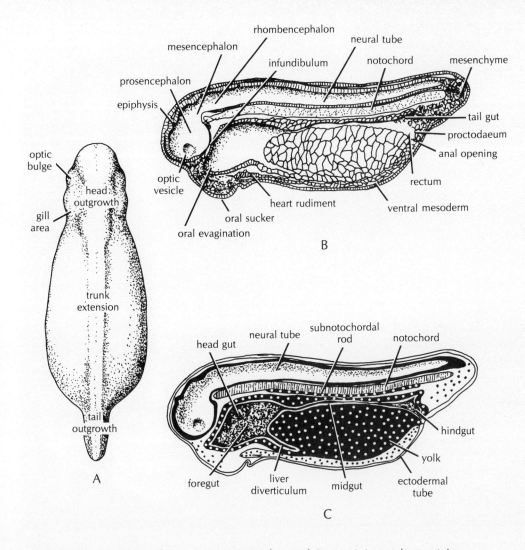

Figure 29. Structure of 3½- to 4-mm. embryo of *Rana pipiens* (about eight pairs of somites are present). (A) External dorsal view. (B) Midsagittal view. (C) Same, showing major organ-forming areas. From O. E. Nelsen, *Comparative Embryology of the Vertebrates* (New York: McGraw-Hill, 1953).

E. Frog, 5–8 mm Stages

tail fin   somites

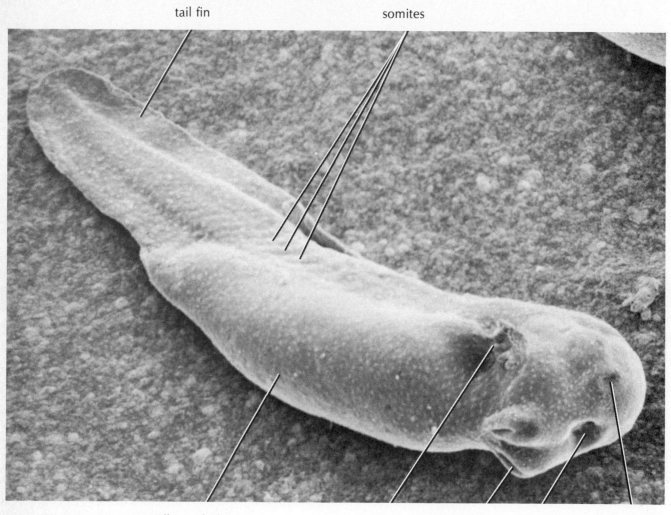

yolky endoderm   gill   adhesive gland   stomodeum   olfactory pit
(below epidermis and mesoderm)

Figure 30. Frog embryo, 5–6 mm, scanning electron micrograph. (55.9×)

developing
brain   eye   pharynx   heart   gill   yolky endoderm

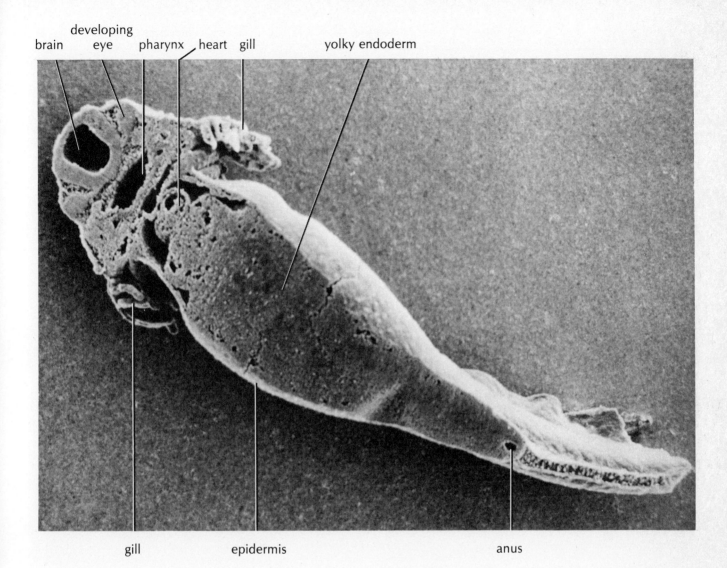

gill   epidermis   anus

Figure 31. Frog embryo, 6—7 mm, horizontal section, scanning electron micrograph. (59.3×)

developing eye   brain   pharynx   heart   gill   epidermis

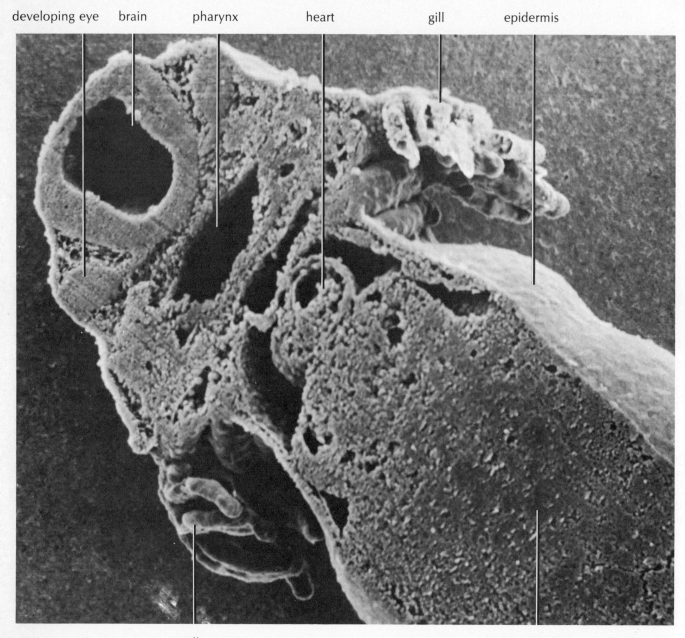

gill   yolky endoderm

Figure 32. Frog embryo, 6–7 mm, horizontal section, scanning electron micrograph. (144.2×)

diencephalon   optic cup   lens vesicle

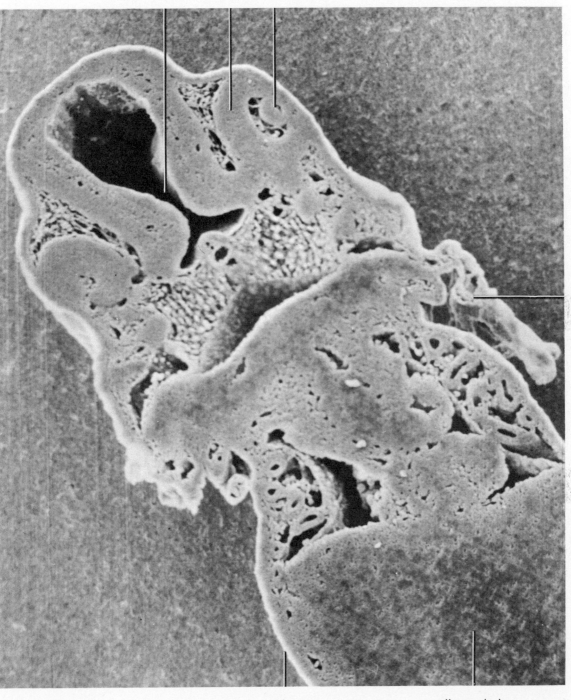

gill

epidermis   yolky endoderm

Figure 33. Frog embryo, 6–7 mm, horizontal section, scanning electron micrograph. (146.9×)

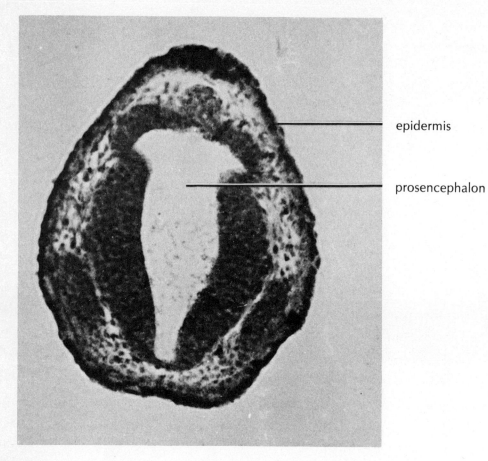

epidermis

prosencephalon

Figure 34. Frog embryo, 5–7 mm, transverse section through prosencephalon. (120×)

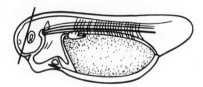

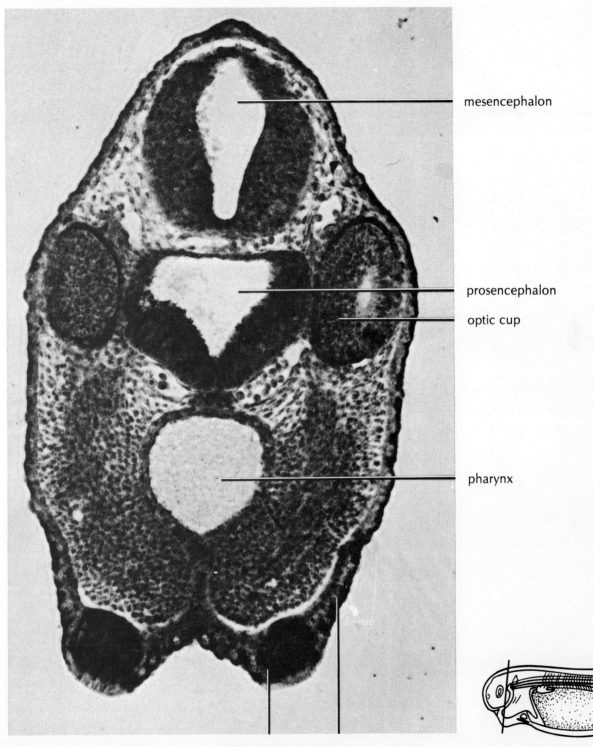

mesencephalon

prosencephalon

optic cup

pharynx

adhesive gland   epidermis

Figure 35.   Frog embryo, 5–7 mm, transverse section through optic cup.
(120×)

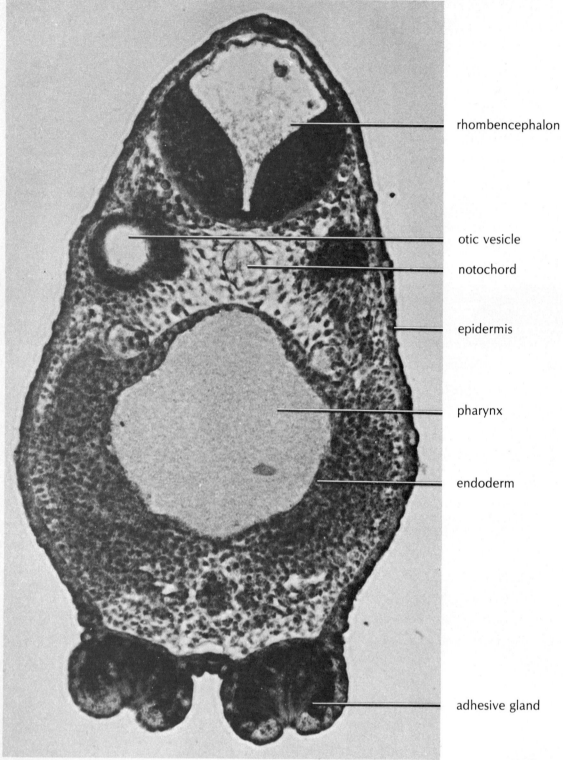

rhombencephalon

otic vesicle

notochord

epidermis

pharynx

endoderm

adhesive gland

Figure 36.   Frog embryo, 5–7 mm, transverse section through otic vesicle. (120×)

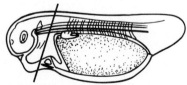

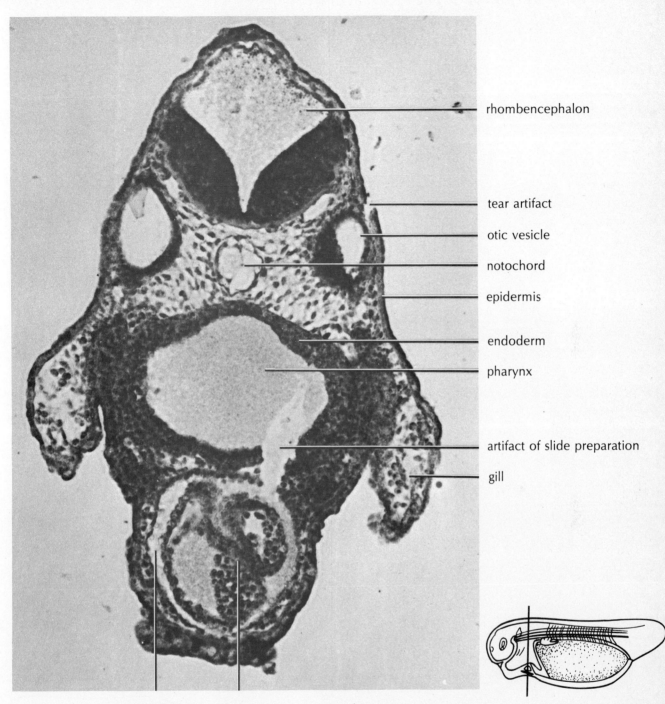

rhombencephalon

tear artifact

otic vesicle

notochord

epidermis

endoderm

pharynx

artifact of slide preparation

gill

pericardial cavity   heart

Figure 37. Frog embryo, 5–7 mm, transverse section through heart. (105.6×)

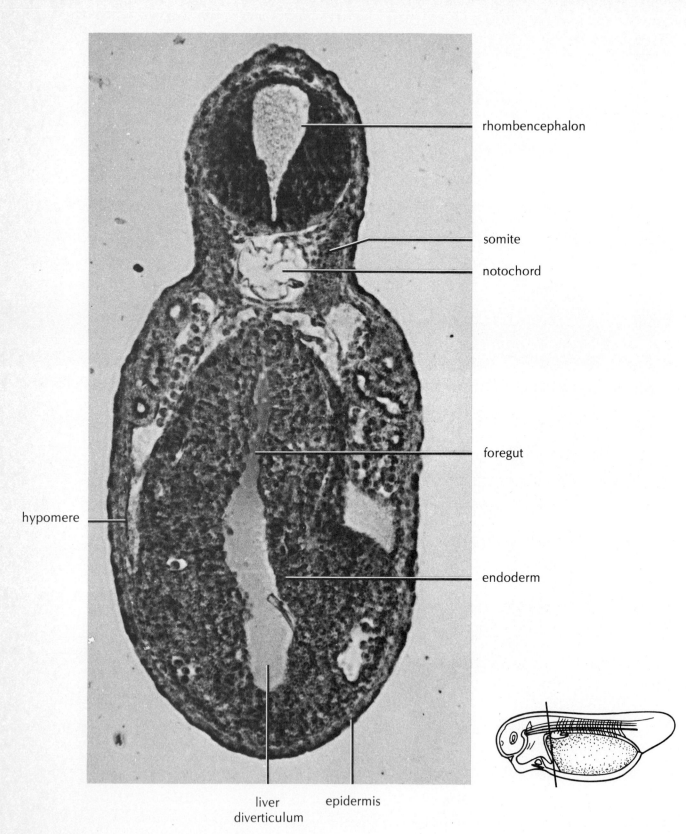

rhombencephalon

somite

notochord

foregut

hypomere

endoderm

liver
diverticulum

epidermis

Figure 38.   Frog embryo, 5–7 mm, transverse section through liver diverticulum. (120×)

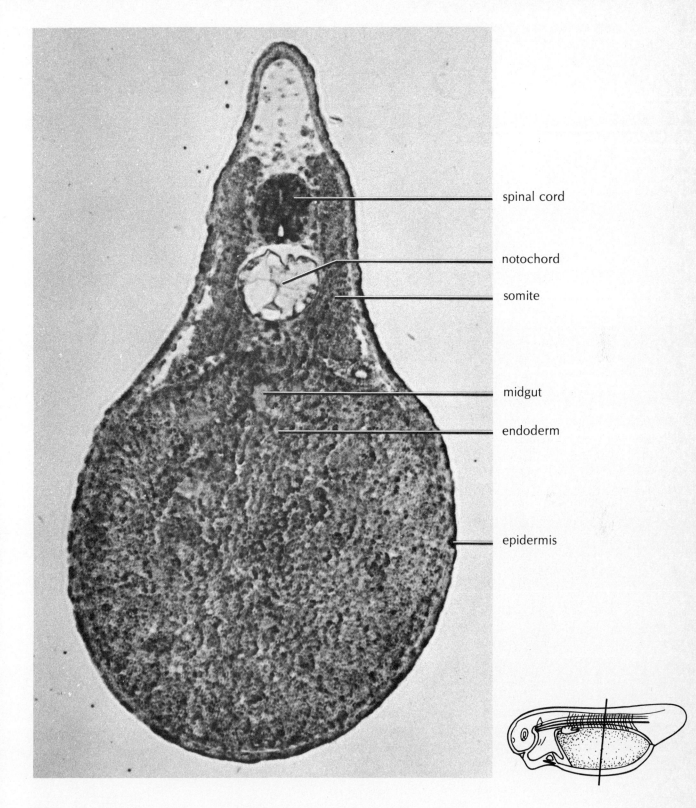

spinal cord

notochord

somite

midgut

endoderm

epidermis

Figure 39.   Frog embryo, 5–7 mm, transverse section through midgut. (120×)

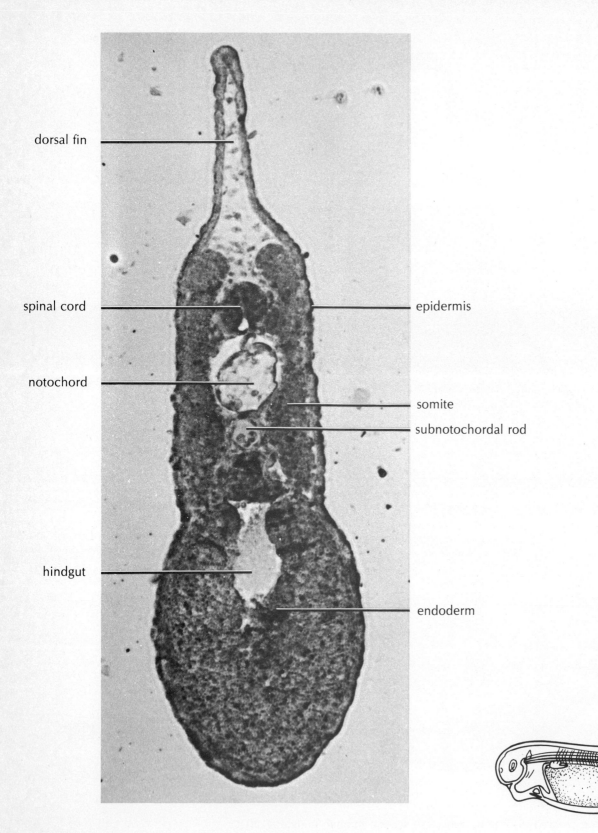

dorsal fin

spinal cord

notochord

hindgut

epidermis

somite

subnotochordal rod

endoderm

Figure 40.   Frog embryo, 5–7 mm, transverse section through hindgut. (120×)

tail fin   gill   olfactory pit

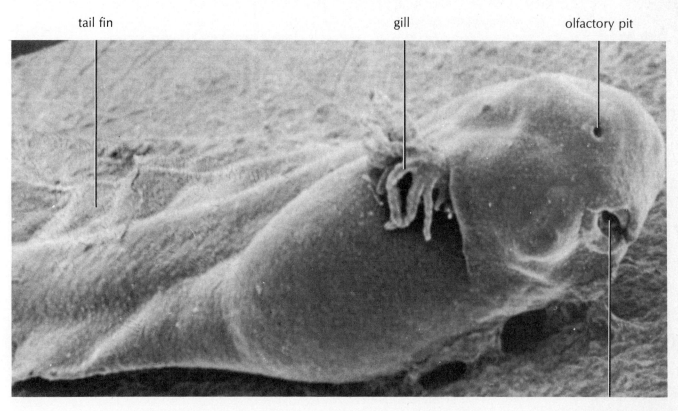

stomodeum

Figure 41. Frog embryo, 7–8 mm, scanning electron micrograph. (79×)

stomodeum   olfactory pit

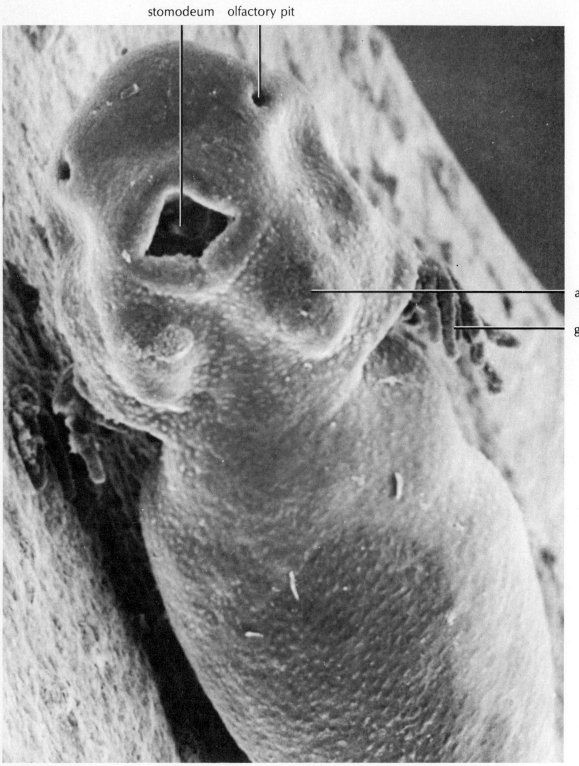

adhesive gland

gill

Figure 42. Frog embryo, 7–8 mm, scanning electron micrograph. (130×)

prosencephalon

mesencephalon

mouth   rhombencephalon

epidermis

somites

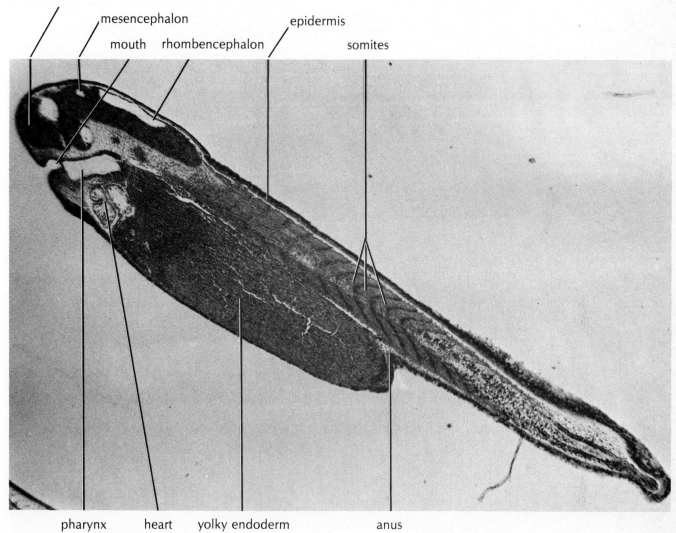

pharynx   heart   yolky endoderm   anus

Figure 43. Frog embryo, 7–8 mm, sagittal section. (32.8×)

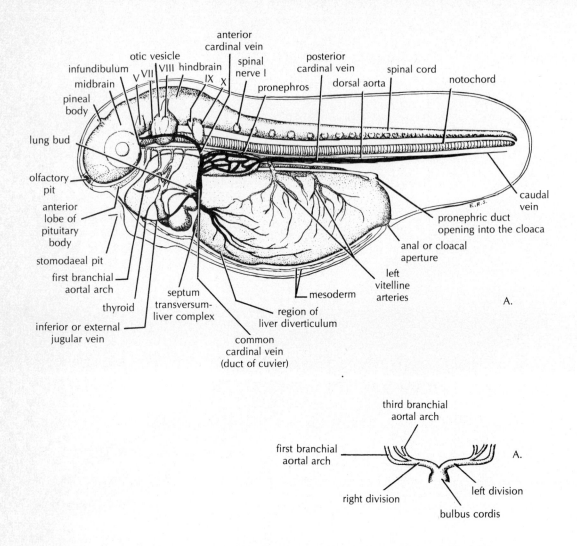

Figure 44. Early frog tadpole, sketch from O. E. Nelsen, *Comparative Embryology of the Vertebrates*, McGraw Hill, 1953.

F. Frog, 10–18 mm Stages

mouth   heart

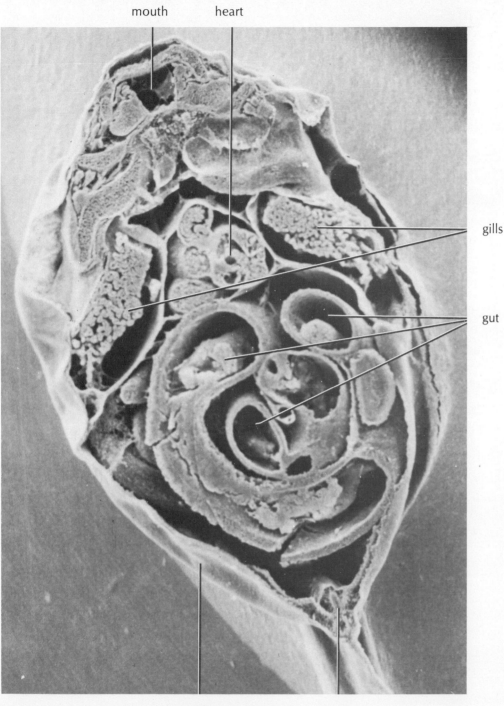

gills

gut

epidermis   anus

Figure 45. Frog tadpole, 11 mm, horizontal section, scanning electron micrograph. (52.3×)

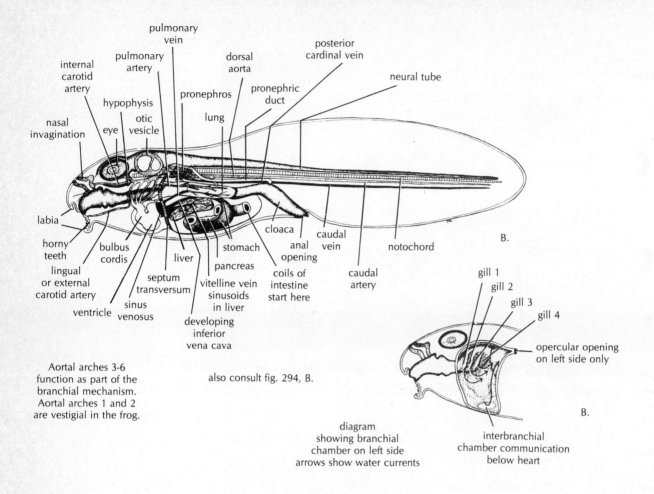

internal
carotid
artery

pulmonary
vein

pulmonary
artery

dorsal
aorta

pronephros

posterior
cardinal vein

pronephric
duct

neural tube

hypophysis

nasal
invagination

eye

otic
vesicle

lung

labia

horny
teeth

bulbus
cordis

lingual
or external
carotid artery

septum
transversum

liver

ventricle

sinus
venosus

vitelline vein
sinusoids
in liver

developing
inferior
vena cava

pancreas

stomach

cloaca

anal
opening

coils of
intestine
start here

caudal
vein

notochord

caudal
artery

B.

Aortal arches 3-6
function as part of the
branchial mechanism.
Aortal arches 1 and 2
are vestigial in the frog.

also consult fig. 294, B.

gill 1

gill 2

gill 3

gill 4

opercular opening
on left side only

interbranchial
chamber communication
below heart

B.

diagram
showing branchial
chamber on left side
arrows show water currents

Figure 46.   Frog tadpole, 10–18 mm, anatomy. From O. E. Nelsen, *Comparative Embryology of the Vertebrates* (New York: McGraw-Hill, 1953).

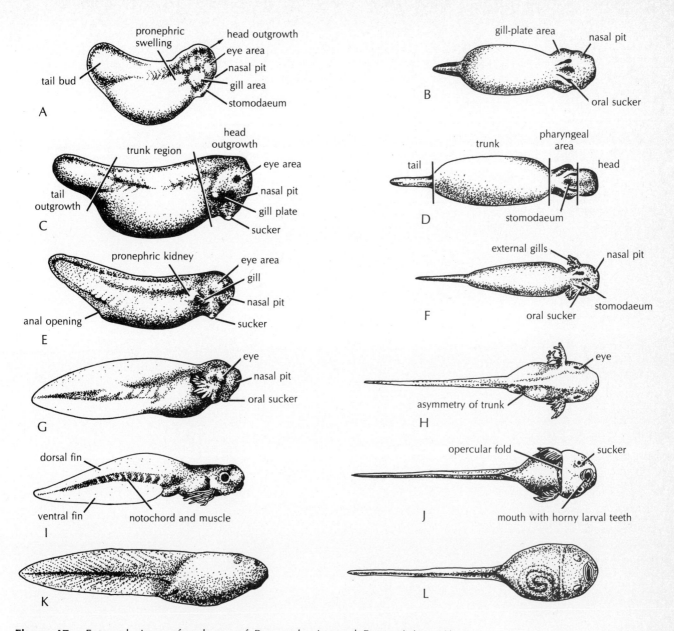

Figure 47. External views of embryos of *Rana sylvatica* and *Rana pipiens*. (A to J after Pollister and Moore: Anat. Rec., 68; K and L after Shumway: Anat. Rec., 78.) (A, B) Lateral and ventral views of 5-mm. stage. Muscular movement is evident at this stage, expressed by simple unilateral flexure; tail is about one-fifth body length. (Pollister and Moore, stage 18.) (C, D) Lateral and ventral views of 6-mm. stage. Primitive heart has developed and begins to beat; tail equals one-third length of body. (Pollister and Moore, stage 19). (E, F) Similar views of 7-mm. stage. Gill circulation is established; hatches; swims; tail equals one-half length of body. (Pollister and Moore, stage 20.) (G, H) Ten-mm. stage, lateral and dorsal views. Gills elongate; tail fin is well developed and circulation is established within; trunk is asymmetrical coincident with posterior bend in the gut tube; cornea of eyes is transparent; epidermis is becoming transparent. (Pollister and Moore, stage 22.) (I, J) Eleven-mm. stage, true tadpole shape. Opercular fold is beginning to develop and gradually growing back over gills. (K, L) Eleven-mm. stage of *R. pipiens* embryo. Observe that opercular folds have grown back over external gills and developing limb buds; opercular chamber opens on left side of body only. From O. E. Nelsen, *Comparative Embryology of the Vertebrates* (New York: McGraw-Hill, 1953).

II
The Chick

Table 2. Chick Development

Hours	Characteristics
Freshly laid	prestreak; embryonic shield forming
6	initial primitive streak
12	primitive streak extending
18	definitive primitive streak; primitive groove and Hensen's node present
19	head process; early notochord extends beyond Hensen's node
20	head fold
22	neural plate begins to fold; 2 somites
24	neural folds better developed; 4 somites
27	endocardial and myocardial tubes forming; enlarged forebrain; 7 somites
30	heart tubes fuse to form a single tube; heart ventricle begins to beat; forebrain, midbrain, and hindbrain visible; 10 somites
33	heart tube bent to right; heart ventricle and atrium beat; first pair of aortic arches present; optic vesicles well defined; 13 somites
36	well developed heartbeat; auditory placodes present; pronephric tubules first appear; 16 somites
40	head torsion begins; hindbrain divided; lens placodes form; 18 somites
44	liver primordium begins to appear; 22 somites
48	sinus venosus develops; 25 somites
54	two pairs of aortic arches present, third forming; hindgut present; 29 somites
60	fourth pair of aortic arches beginning to form; 31 somites
72	telencephalic vesicles present; dorsal pancreas evagination appears; 35 somites

The student will quickly see that there is variability in the developmental progress within a group of eggs incubated for a specific time. This is due to many factors, including varying amounts of incubation before removal from the nest to begin incubation in the laboratory. This table should be useful to the student because it includes some of the major characteristics that appear at approximate times during chick development. The hours of development and the characteristics given here are approximations. Slight differences will appear in other similar charts or tables. The table was compiled based upon the work of: Bradley M. Patten, *Early Embryology of the Chick* (New York: McGraw Hill, 1971), and V. Hamburger and H. L. Hamilton, A series of normal stages in the development of the chick embryo, *Journal of Morphology,* 88: 49–92, 1951.

A. Chick Gastrula

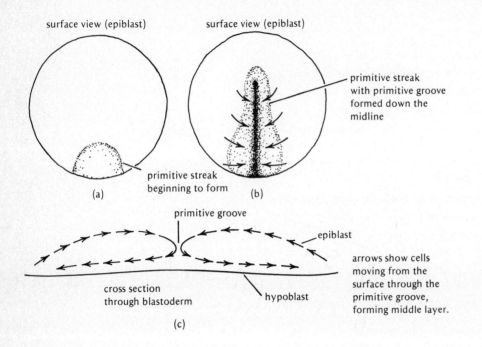

surface view (epiblast) surface view (epiblast)

primitive streak
with primitive groove
formed down the
midline

primitive streak
beginning to form

(a) (b)

primitive groove

epiblast

arrows show cells
moving from the
surface through the
primitive groove,
forming middle layer.

cross section
through blastoderm

hypoblast

(c)

Figure 48. Bird gastrulation. From S. B. Oppenheimer, *Introduction to Embryonic Development,* 2nd Ed. (Boston: Allyn and Bacon, 1984).

notochordal process   neural plate   proamnion

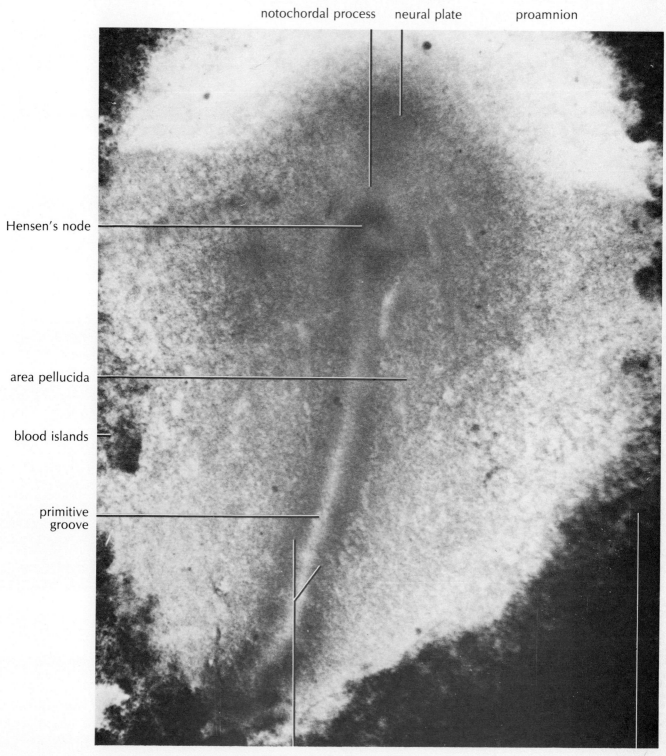

Hensen's node

area pellucida

blood islands

primitive
groove

primitive folds   area opaca

Figure 49.   Chick embryo whole mount, 18 hours. (85.1×)

Hensen's node   primitive folds   area opaca

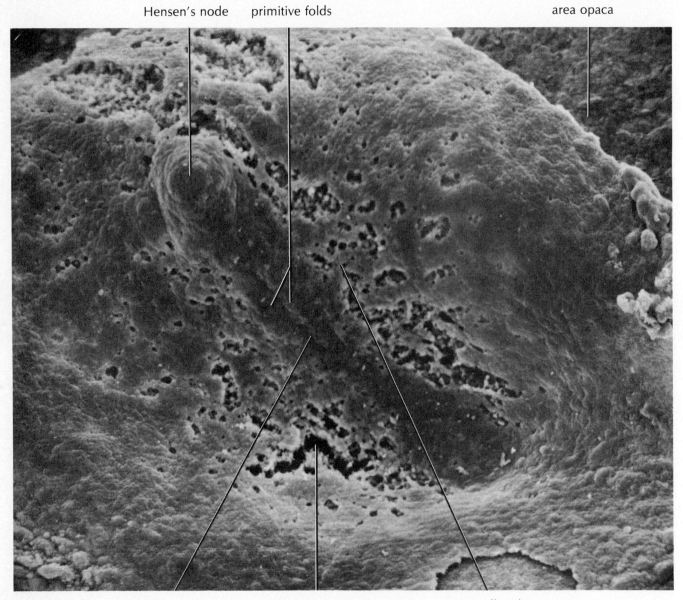

primitive groove   subgerminal space   area pellucida
                    (below blastoderm in area pellucida)

Figure 50. Chick embryo, 16–20 hours, scanning electron micrograph.
(161.9×)

B. Chick, 26–33 Hour Embryos

proamnion   anterior neuropore

head ectoderm

head mesenchyme

area opaca

neural tube

anterior intestinal portal

neural fold

intersomitic groove

somite

area pellucida

segmental mesoderm

notochord

primitive knot (Hensen's node)

primitive streak

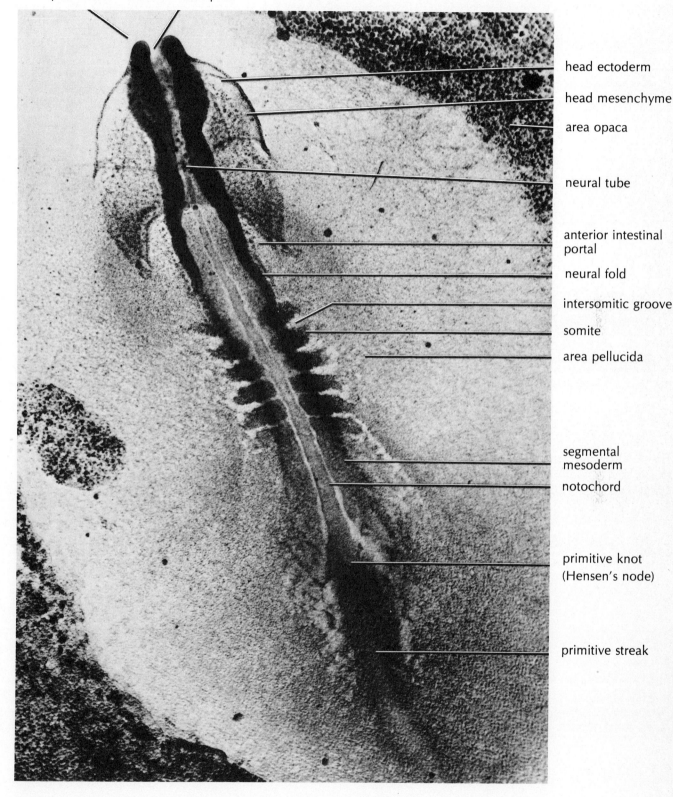

Figure 51. Chick embryo whole mount, 26–29 hours. (65.7×)

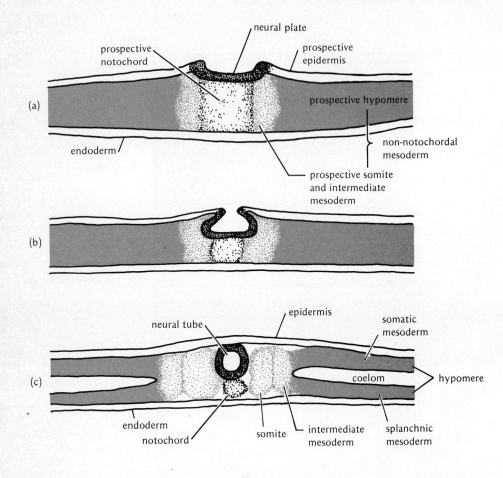

Figure 52. Mesodermal separation in chick embryo. From S. B. Oppenheimer, *Introduction to Embryonic Development,* 2nd Ed. (Boston: Allyn and Bacon, 1984).

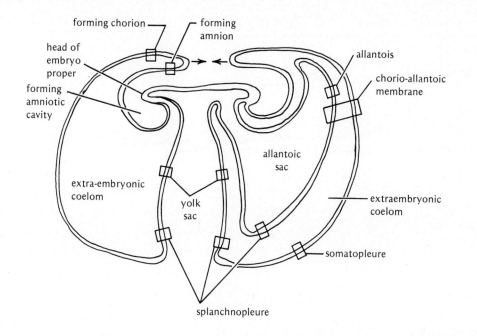

Figure 53. Summary diagram of early chick embryo with its forming extraembryonic membranes. Longitudinal view. From S. B. Oppenheimer, *Introduction to Embryonic Development,* 2nd Ed. (Boston: Allyn and Bacon, 1984).

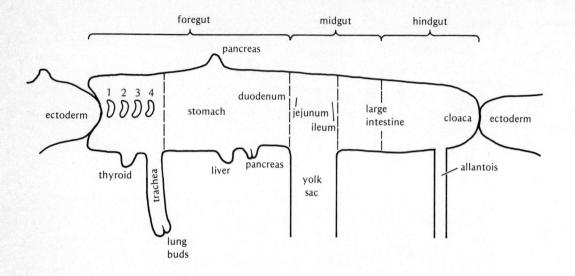

Figure 54. Endodermal derivatives. Diagram of the gut tube and its outpocketings. From S. B. Oppenheimer, *Introduction to Embryonic Development,* 2nd Ed. (Boston: Allyn and Bacon, 1984).

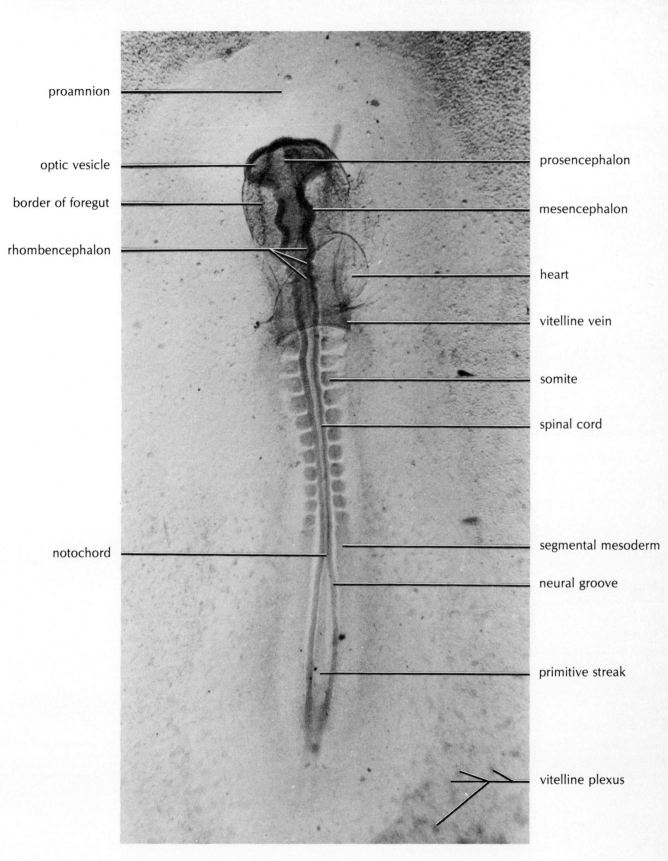

proamnion

optic vesicle

border of foregut

rhombencephalon

notochord

prosencephalon

mesencephalon

heart

vitelline vein

somite

spinal cord

segmental mesoderm

neural groove

primitive streak

vitelline plexus

Figure 55.   Chick embryo, whole mount, 33 hours. (40.5×)

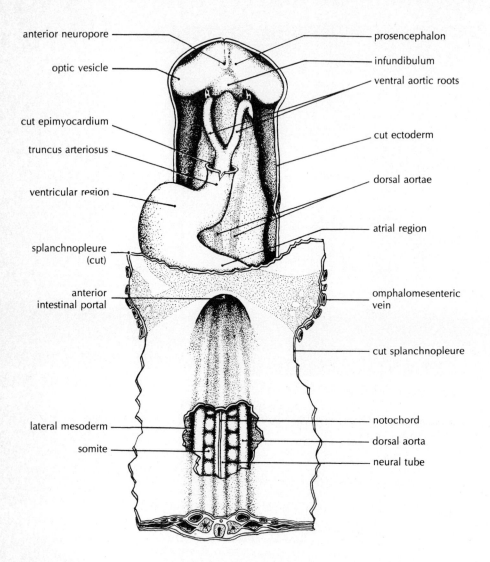

Figure 56. Diagrammatic ventral view of dissection of a 35-hour chick embryo. The splanchnopleure of the yolk-sac cephalic to the anterior intestinal portal, the ectoderm of the ventral surface of the head, and the mesoderm of the pericardial region have been removed to show the underlying structures. *(Modified from Pentiss)* From B. M. Patten, *Early Embryology of the Chick* (New York: McGraw-Hill, 1957).

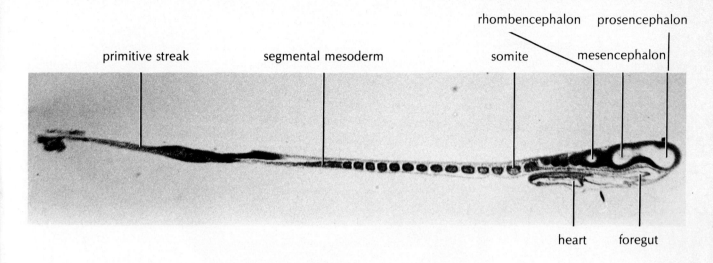

Figure 57. Chick embryo, 33 hours, sagittal section. (32.4×)

somatopleure

ectoderm   splanchnic   prosencephalon   epidermis   prospective
mesoderm   lens   somatopleure

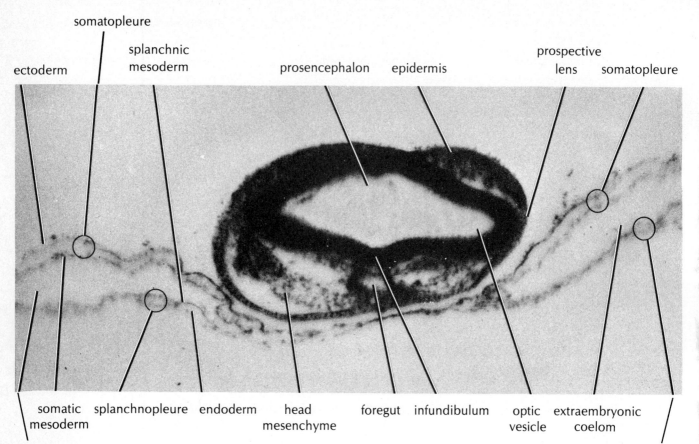

somatic   splanchnopleure   endoderm   head   foregut   infundibulum   optic   extraembryonic
mesoderm   mesenchyme   vesicle   coelom

extraembryonic
coelom

splanchnopleure

Figure 58. Chick embryo, 33 hours, transverse section through prosencephalon. (164.7×)

head epidermis   notochord

prosencephalon

optic vesicle

prospective lens

dorsal aorta

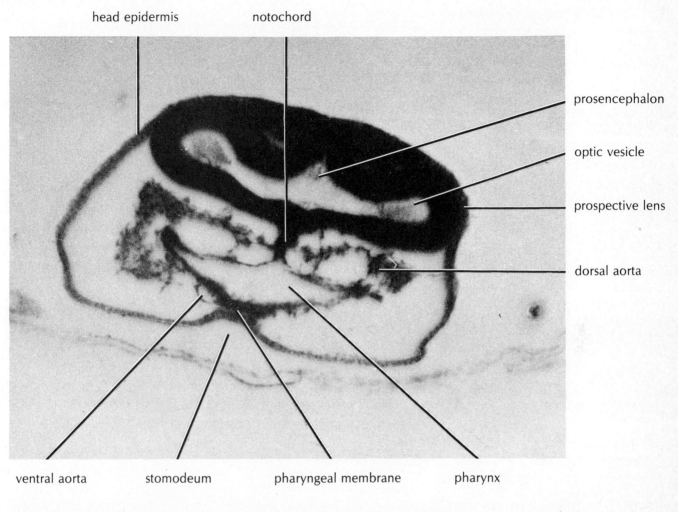

ventral aorta   stomodeum   pharyngeal membrane   pharynx

Figure 59. Chick embryo, 33 hours, transverse section through optic vesicles. (185×)

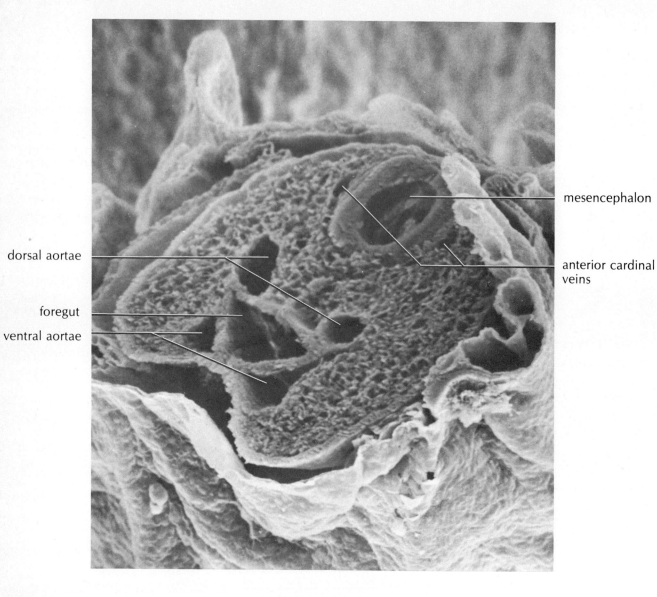

dorsal aortae

foregut

ventral aortae

mesencephalon

anterior cardinal veins

Figure 60.   Chick embryo, 33 hours, transverse section through mesencephalon, scanning electron micrograph. (213×)

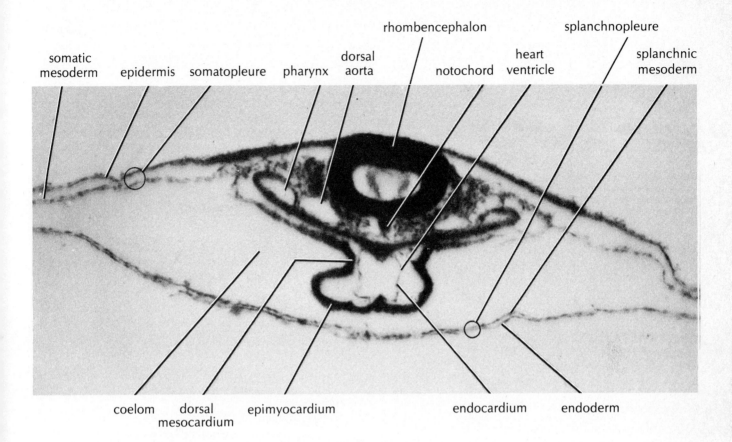

somatic mesoderm   epidermis   somatopleure   pharynx   dorsal aorta   rhombencephalon   notochord   heart ventricle   splanchnopleure   splanchnic mesoderm

coelom   dorsal mesocardium   epimyocardium   endocardium   endoderm

Figure 61.   Chick embryo, 33 hours, transverse section through heart. (159.1×)

rhombencephalon   extraembryonic coelom

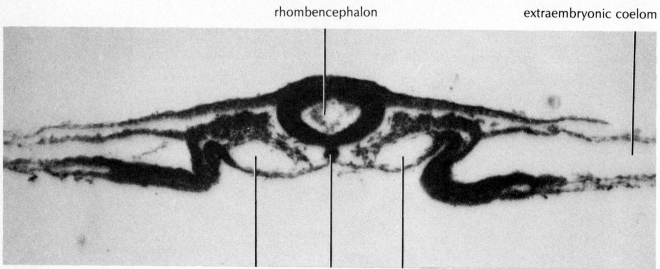

dorsal aorta   notochord   dorsal aorta

Figure 62. Chick embryo, 33 hours, transverse section through rhombencephalon. (159.1×)

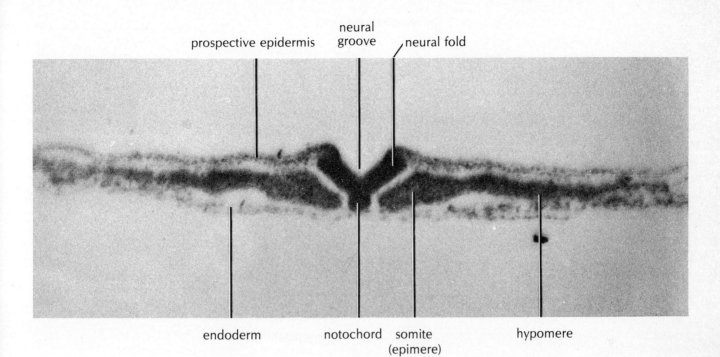

Figure 63. Chick embryo, 33 hours, transverse section through somites. (179.5×)

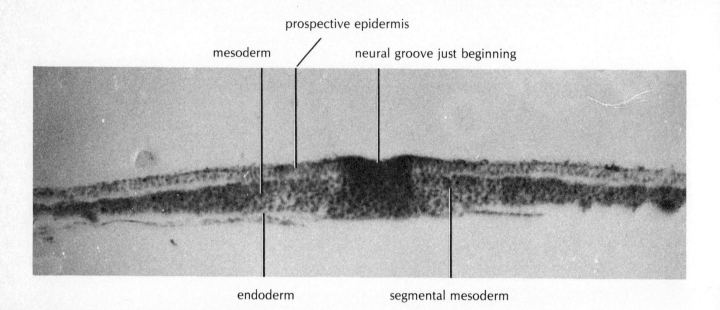

Figure 64. Chick embryo, 33 hours, transverse section through segmental mesoderm. (170.2×)

C. Chick, 50 Hour Embryo

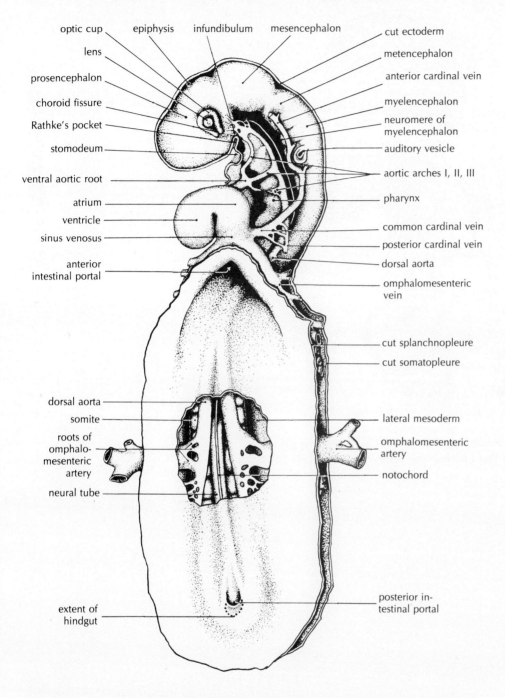

optic cup
epiphysis
infundibulum
mesencephalon
cut ectoderm
lens
metencephalon
prosencephalon
anterior cardinal vein
choroid fissure
myelencephalon
Rathke's pocket
neuromere of myelencephalon
stomodeum
auditory vesicle
ventral aortic root
aortic arches I, II, III
atrium
pharynx
ventricle
sinus venosus
common cardinal vein
posterior cardinal vein
anterior intestinal portal
dorsal aorta
omphalomesenteric vein
cut splanchnopleure
cut somatopleure
dorsal aorta
somite
lateral mesoderm
roots of omphalo-mesenteric artery
omphalomesenteric artery
notochord
neural tube
extent of hindgut
posterior intestinal portal

Figure 65. Diagram of dissection of chick of about 50 hours. The splanchnopleure of the yolk-sac cephalic to the anterior intestinal portal, the ectoderm of the left side of the head, and the mesoderm in the pericardial region have been dissected away. A window has been cut in the splanchnopleure of the dorsal wall of the midgut to show the origin of the omphalomesenteric arteries. *(Modified from Prentiss.)* From B. M. Patten, *Early Embryology of the Chick* (New York: McGraw-Hill, 1957).

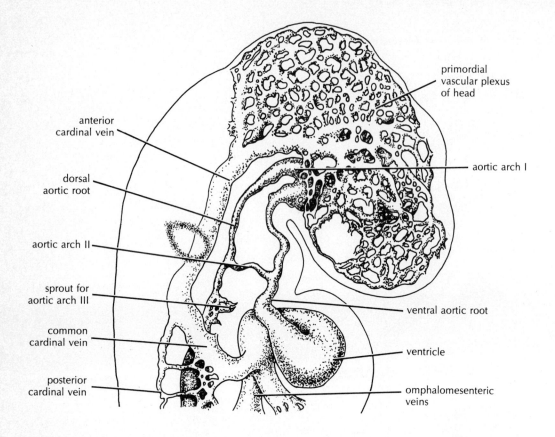

primordial
vascular plexus
of head

anterior
cardinal vein

aortic arch I

dorsal
aortic root

aortic arch II

sprout for
aortic arch III

ventral aortic root

common
cardinal vein

ventricle

posterior
cardinal vein

omphalomesenteric
veins

Figure 66.   Chick embryo, 50 hours, vascular system in anterior region, from B. M. Patten, *Early Embryology of the Chick* (New York: McGraw-Hill, 1957).

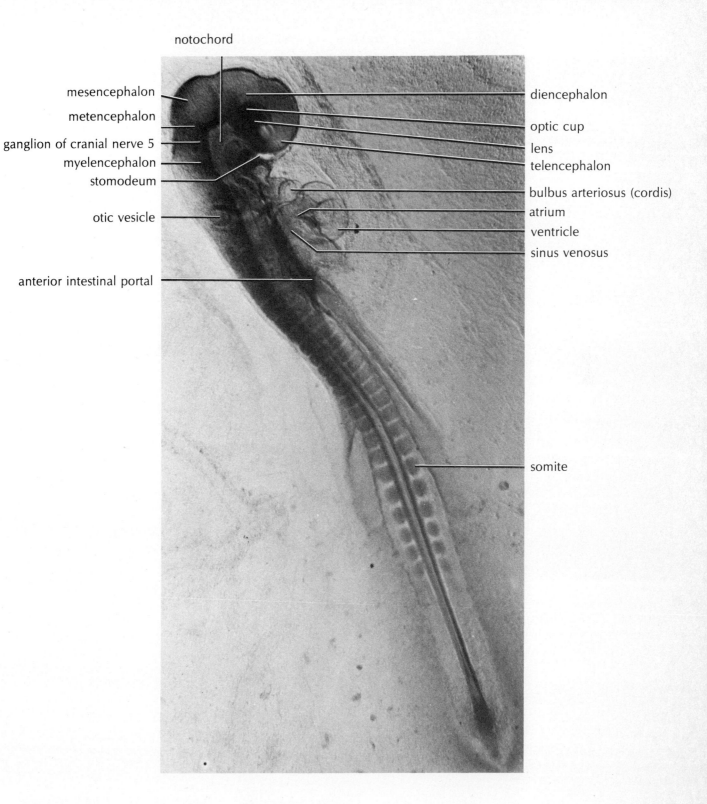

notochord

mesencephalon

metencephalon

ganglion of cranial nerve 5

myelencephalon

stomodeum

otic vesicle

anterior intestinal portal

diencephalon

optic cup

lens

telencephalon

bulbus arteriosus (cordis)

atrium

ventricle

sinus venosus

somite

Figure 67. Chick embryo, 50 hours, whole mount. (32×)

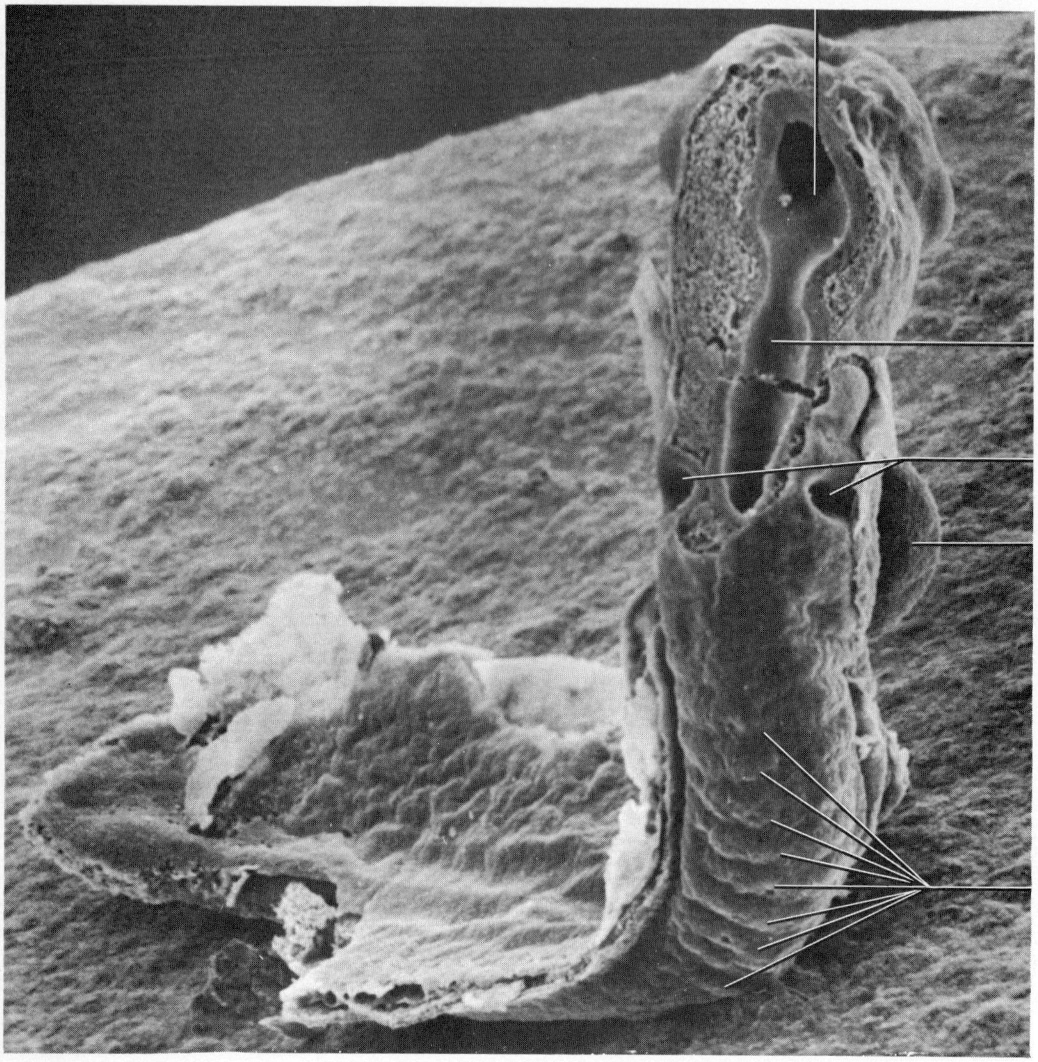

mesencephalon

rhombencephalon

otic pits

heart ventricle

somites

Figure 68.   Chick embryo, about 50 hours, scanning electron micrograph, whole embryo with horizontal section in head region. (75.5×)

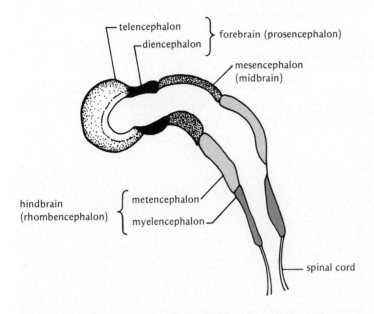

Figure 69. Divisions of the vertebrate brain. From S. B. Oppenheimer, *Introduction to Embryonic Development,* 2nd Ed. (Boston: Allyn and Bacon, 1984).

mesencephalon

rhombencephalon

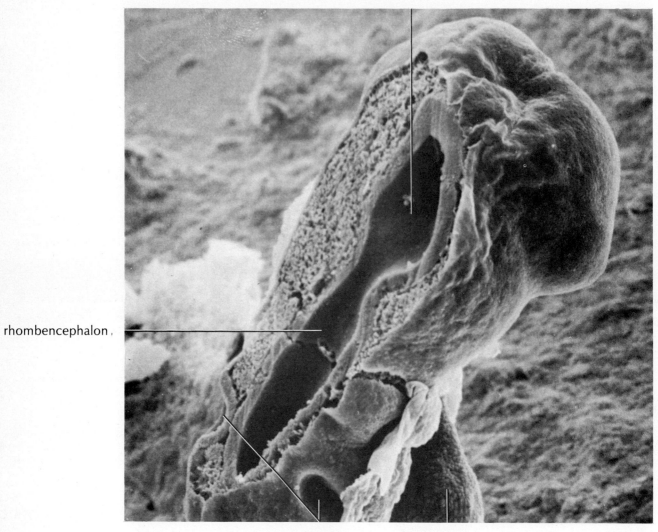

otic pits   heart ventricle

Figure 70. Chick embryo, about 50 hours, horizontal section of head, scanning electron micrograph. (154.2×)

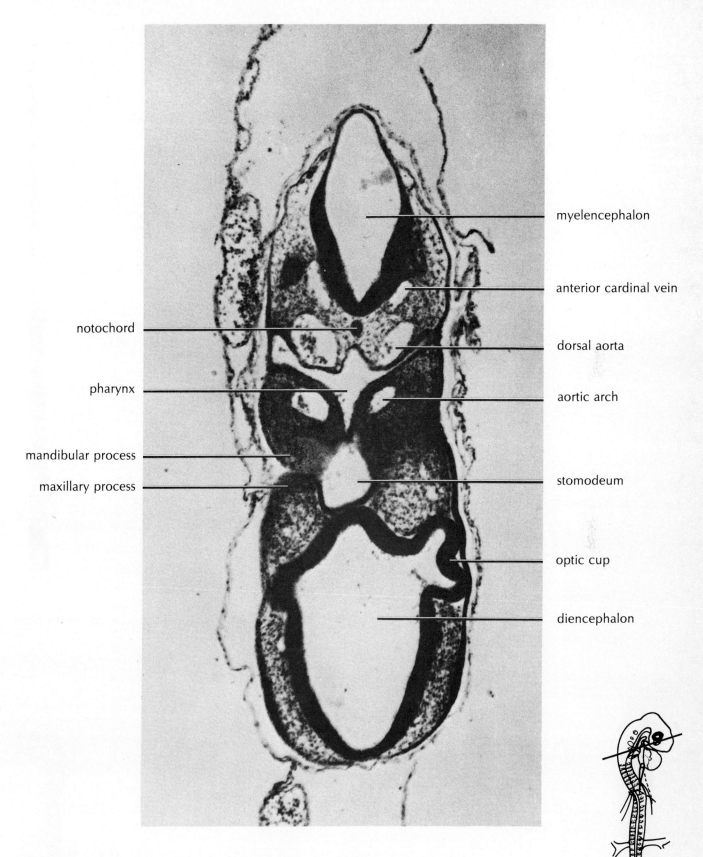

myelencephalon

anterior cardinal vein

notochord

dorsal aorta

pharynx

aortic arch

mandibular process

maxillary process

stomodeum

optic cup

diencephalon

Figure 71. Chick embryo, 50 hours, transverse section through optic cup. (114.7×)

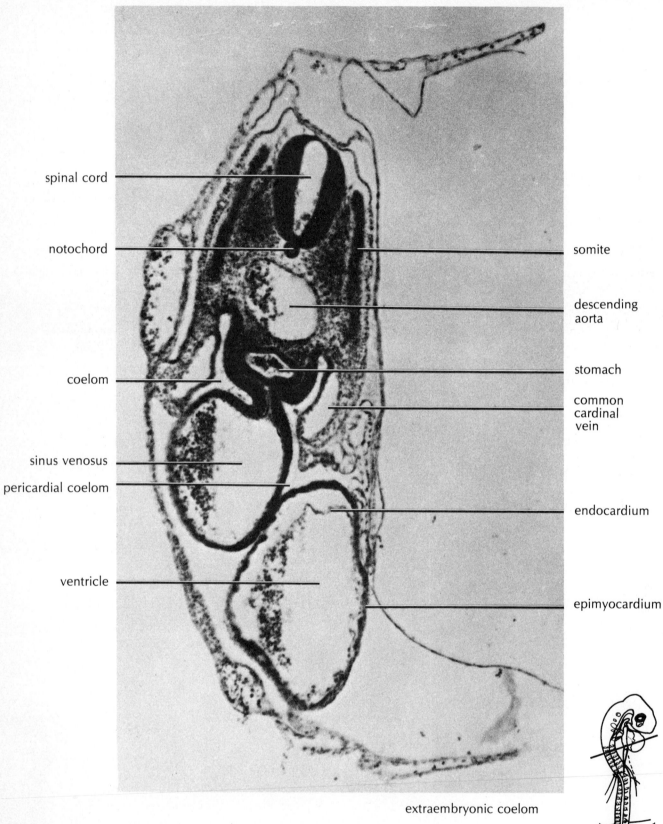

spinal cord

notochord

coelom

sinus venosus

pericardial coelom

ventricle

somite

descending
aorta

stomach

common
cardinal
vein

endocardium

epimyocardium

extraembryonic coelom

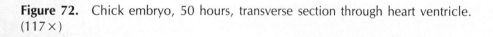

Figure 72. Chick embryo, 50 hours, transverse section through heart ventricle.
(117×)

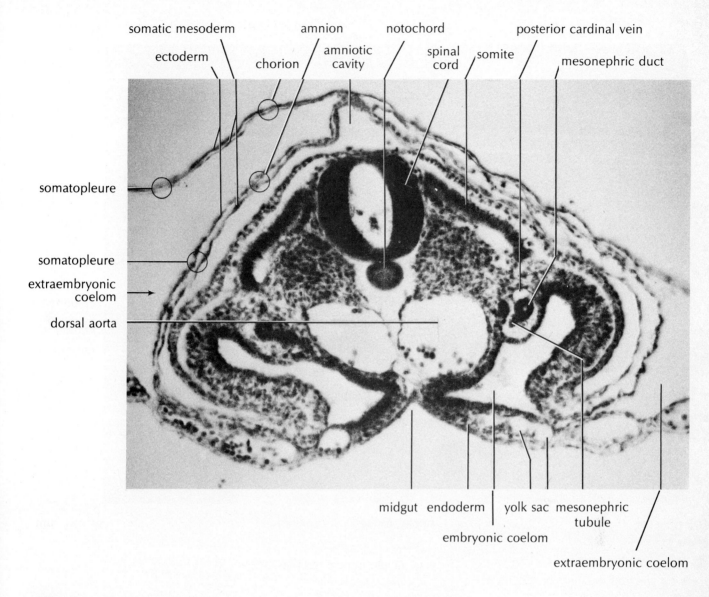

somatic mesoderm   amnion   notochord   posterior cardinal vein

ectoderm   amniotic spinal somite   mesonephric duct

chorion cavity   cord

somatopleure

somatopleure

extraembryonic
coelom

dorsal aorta

midgut endoderm | yolk sac mesonephric
tubule

embryonic coelom

extraembryonic coelom

Figure 73. Chick embryo, 50 hours, transverse section through mesonephric
ducts. (200×)

somites   spinal cord

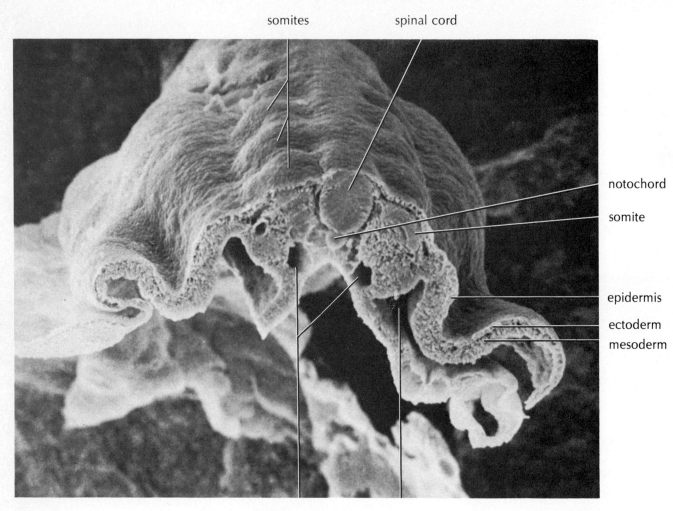

notochord

somite

epidermis

ectoderm

mesoderm

dorsal aortae   coelom

Figure 74.   Chick embryo, 50 hours, transverse section through somites, scanning electron micrograph. (129.6×)

somite (epimere)   spinal cord   epidermis   somatopleure   somatic mesoderm

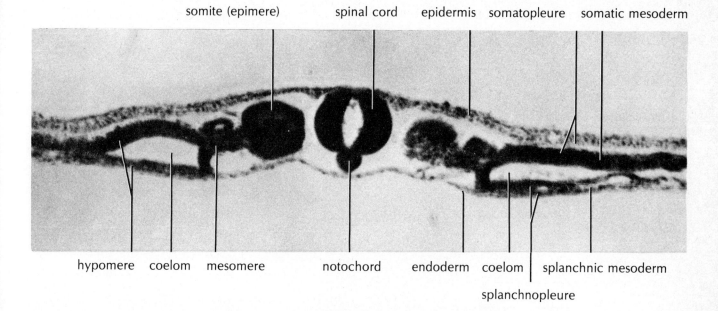

hypomere   coelom   mesomere   notochord   endoderm   coelom | splanchnic mesoderm

splanchnopleure

Figure 75. Chick embryo, 50 hours, transverse section through somites. (148.2×)

D. Chick, 3–4 Day Embryos

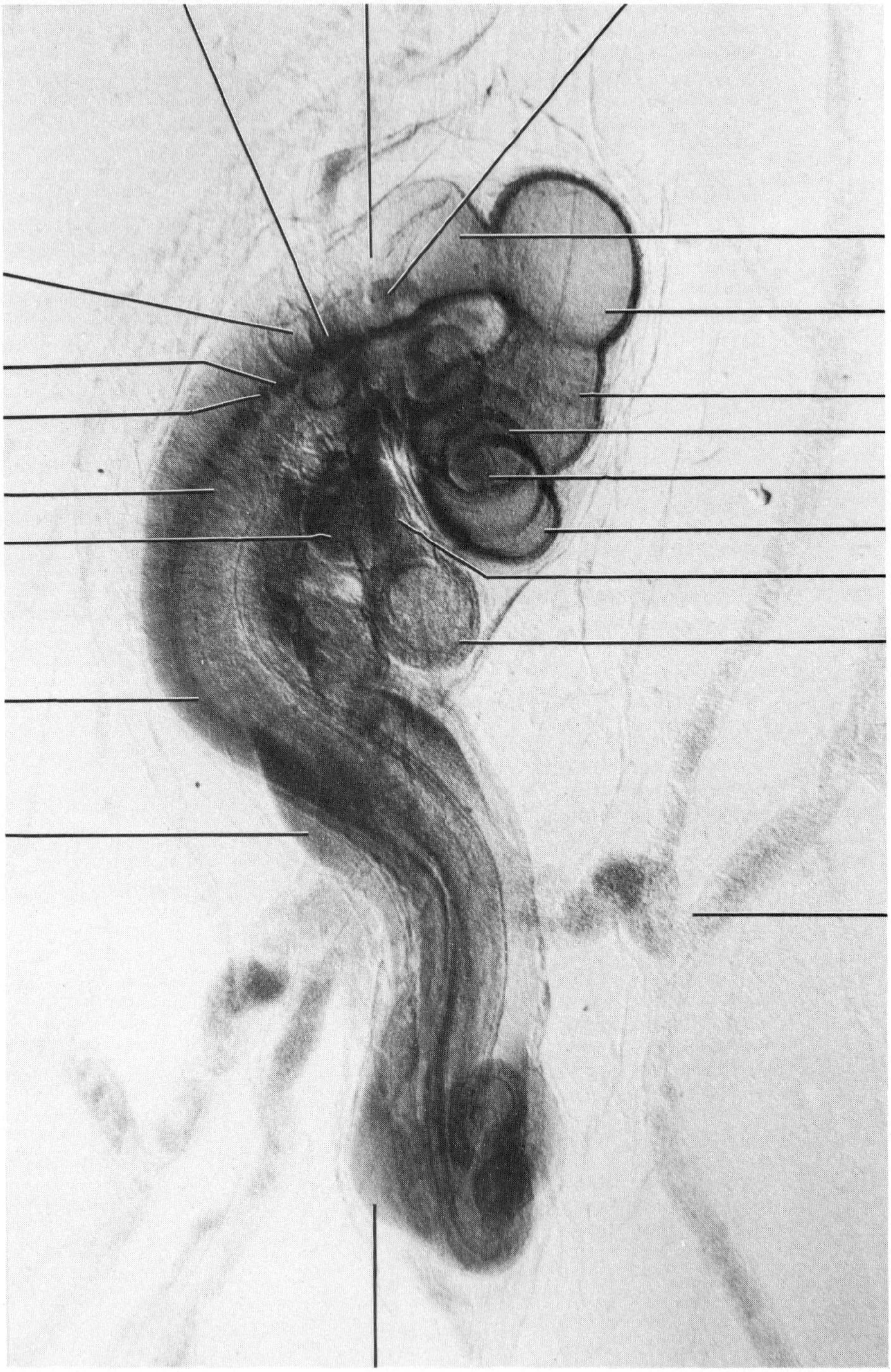

ganglion of cranial nerves 7, 8   myelencephalon   ganglion of cranial nerve

auditory vesicle

ganglion of cranial
nerve 9

ganglion of cranial
nerve 10

somite

atrium

spinal cord

wing bud

metencephalon

mesencephalon

diencephalon

optic cup

lens vesicle

telencephalon

bulbus cordis

ventricle

vitelline vessels

leg bud

Figure 76. Chick embryo, whole mount, 3 days. (19.8×)

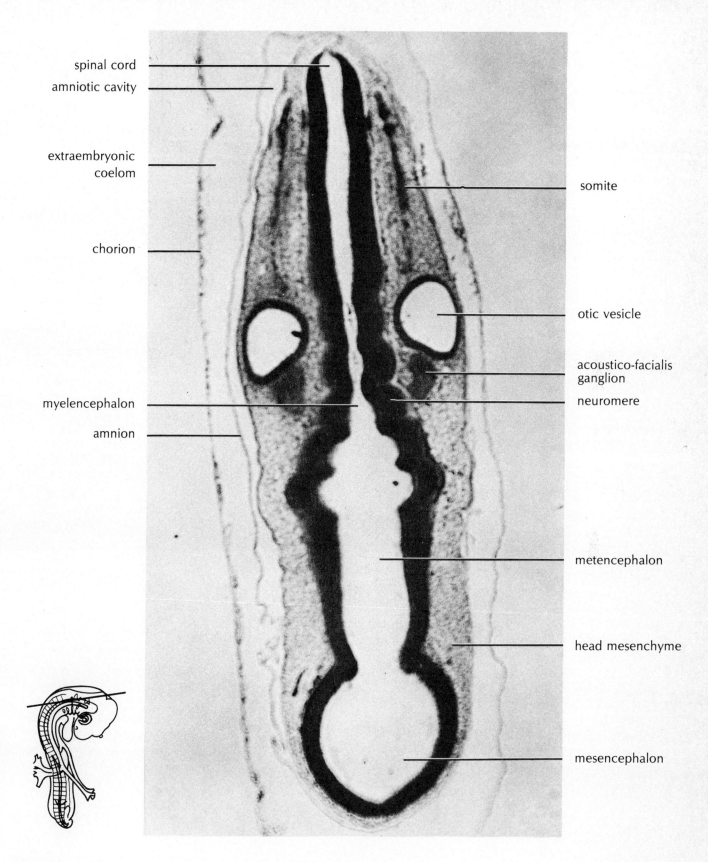

spinal cord

amniotic cavity

extraembryonic coelom

chorion

myelencephalon

amnion

somite

otic vesicle

acoustico-facialis ganglion

neuromere

metencephalon

head mesenchyme

mesencephalon

Figure 77. Chick embryo, 3 days, transverse section through otic vesicles. (64.2×)

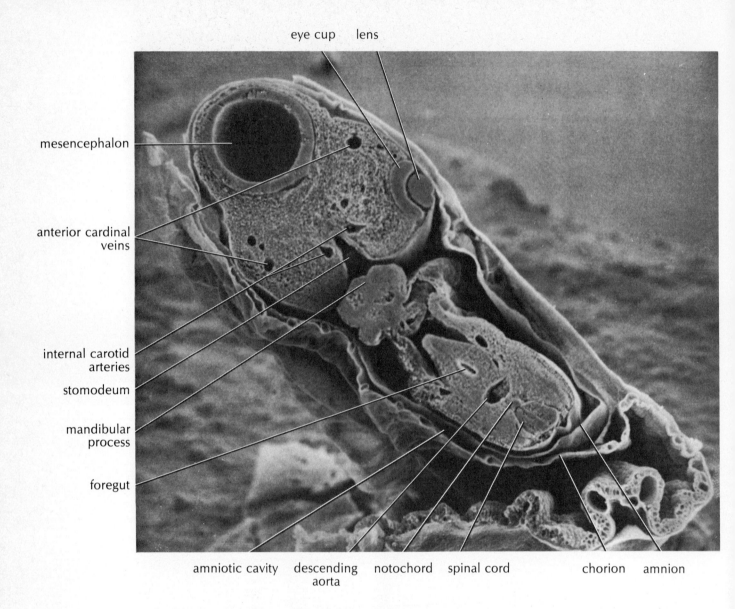

eye cup   lens

mesencephalon

anterior cardinal
veins

internal carotid
arteries

stomodeum

mandibular
process

foregut

amniotic cavity   descending   notochord   spinal cord   chorion   amnion
                   aorta

Figure 78.   Chick embryo, 3 days, transverse section through developing eye,
scanning electron micrograph. (61.2×)

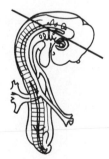

coelom   ductus venosus

spinal cord

notochord

amnion

chorion

descending
aorta

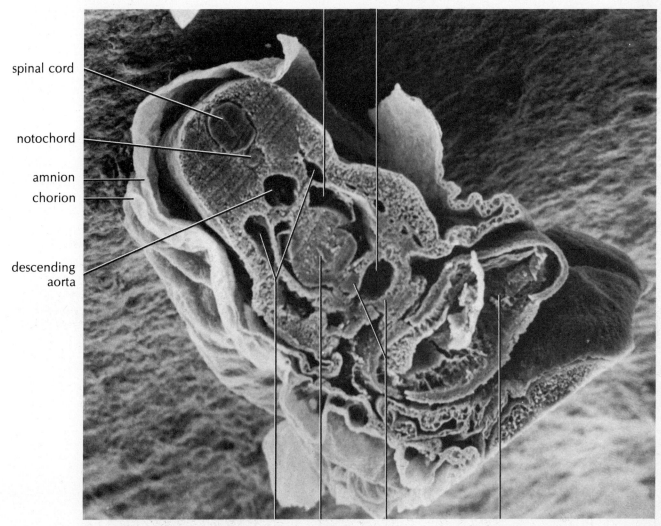

posterior cardinal veins   duodenum   liver   ventricle

Figure 79.   Chick embryo, 3 days, transverse section through heart ventricle,
scanning electron micrograph. (97 ×)

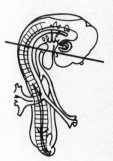

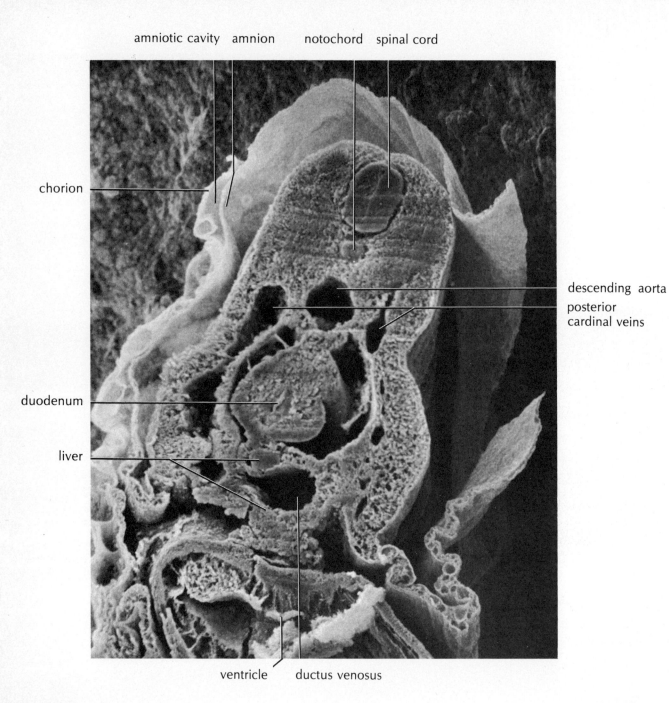

amniotic cavity   amnion   notochord   spinal cord

chorion

descending aorta

posterior
cardinal veins

duodenum

liver

ventricle   ductus venosus

Figure 80. Chick embryo, 3 days, transverse section through ductus venosus, scanning electron micrograph. (145.7×)

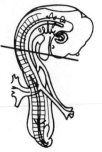

genital ridge   spinal cord

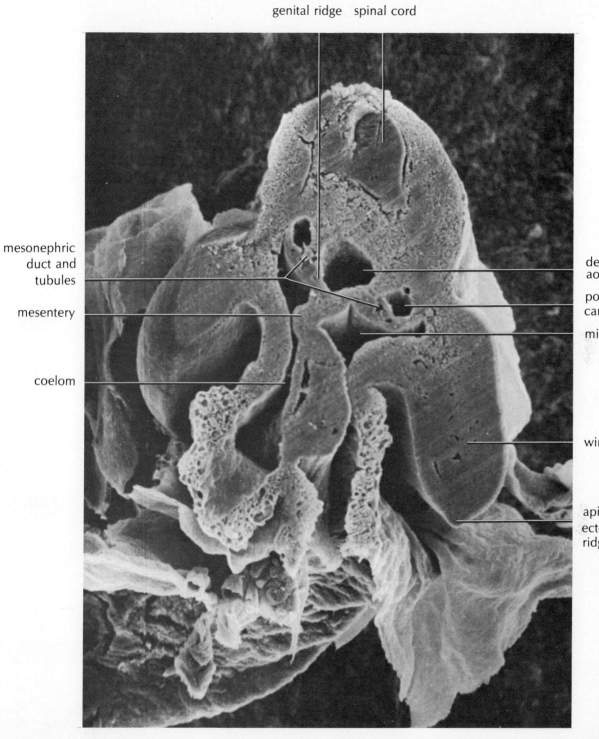

mesonephric
duct and
tubules

mesentery

coelom

descending
aorta

posterior
cardinal vein

midgut

wing bud

apical
ectodermal
ridge

Figure 81. Chick embryo, 3 days, transverse section through wing buds, scanning electron micrograph. (98.6×)

notochord   spinal cord   paired dorsal   mesonephric duct   hindlimb   apical
aortae   bud   ectodermal ridge

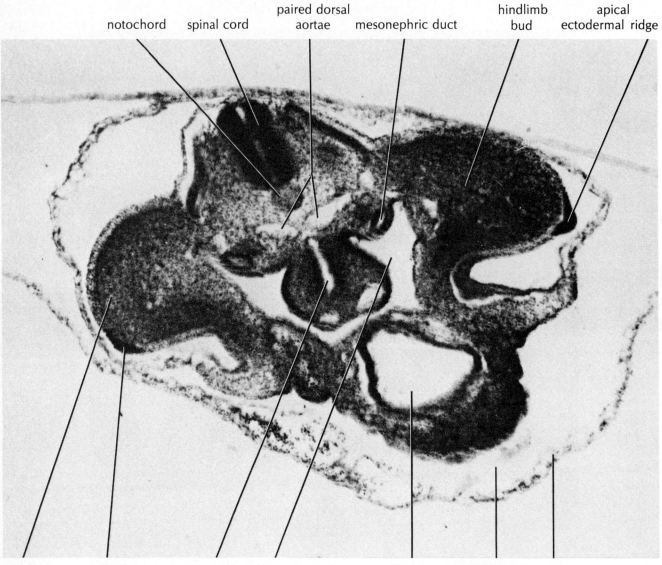

hindlimb   apical   hindgut   embryonic coelom   allantois   extraembryonic   yolk sac
bud   ectodermal ridge   coelom

Figure 82.   Chick embryo, 3 days, transverse section through hindlimb (leg)
buds. (101×)

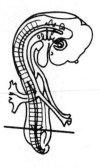

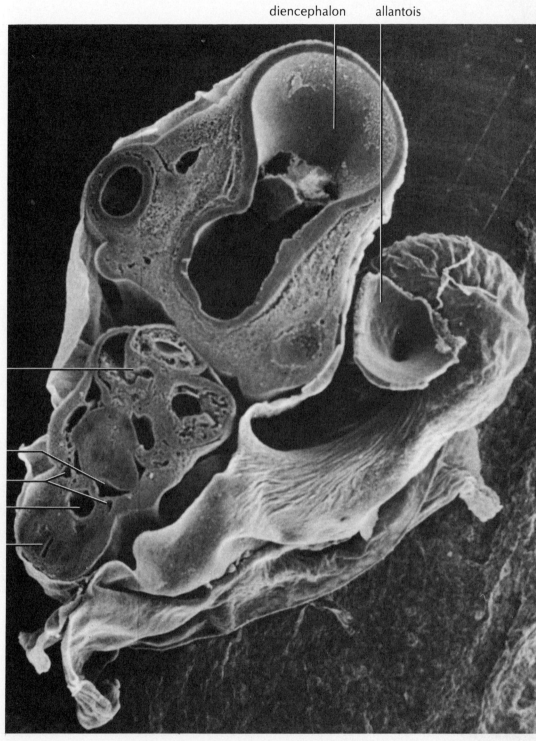

diencephalon   allantois

heart

coelom

posterior
cardinal veins

descending aorta

spinal cord

Figure 83. Chick embryo, 3–4 days, transverse section through brain and spinal cord, scanning electron micrograph. (46.5×)

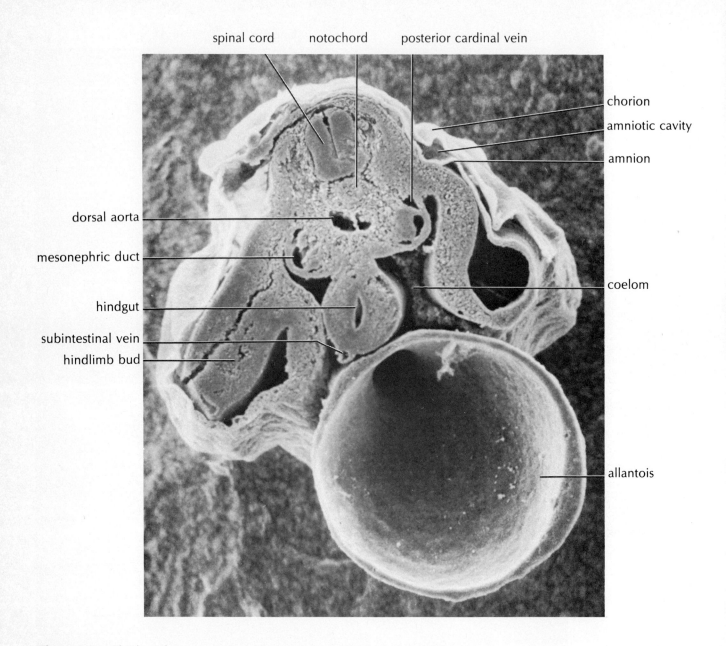

spinal cord   notochord   posterior cardinal vein

chorion
amniotic cavity
amnion

dorsal aorta

mesonephric duct

hindgut

subintestinal vein
hindlimb bud

coelom

allantois

Figure 84. Chick embryo, 3–4 days, transverse section through hindgut and al-lantois, scanning electron micrograph. (81.4×)

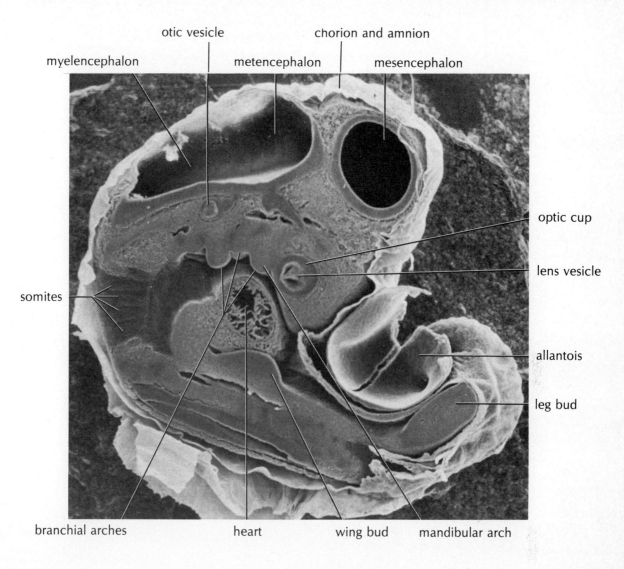

otic vesicle

myelencephalon   metencephalon   chorion and amnion   mesencephalon

optic cup

lens vesicle

somites

allantois

leg bud

branchial arches   heart   wing bud   mandibular arch

Figure 85. Chick embryo, 3–4 days, sagittal section, scanning electron micrograph. (35.2×)

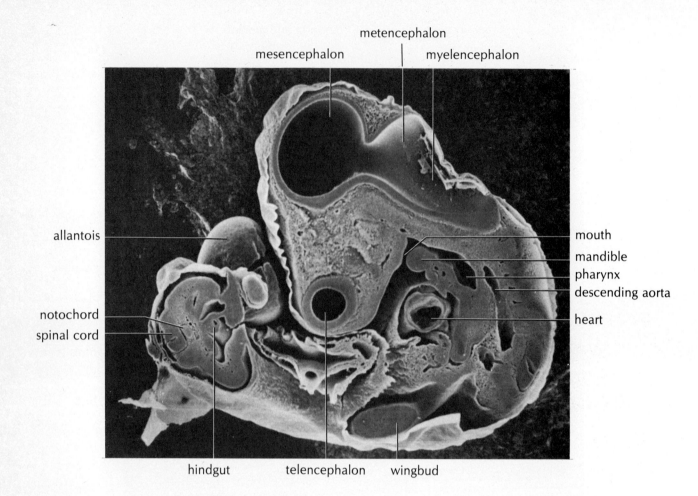

Figure 86. Chick embryo, 3–4 days, sagittal section and transverse section through hindgut, scanning electron micrograph. (31.7×)

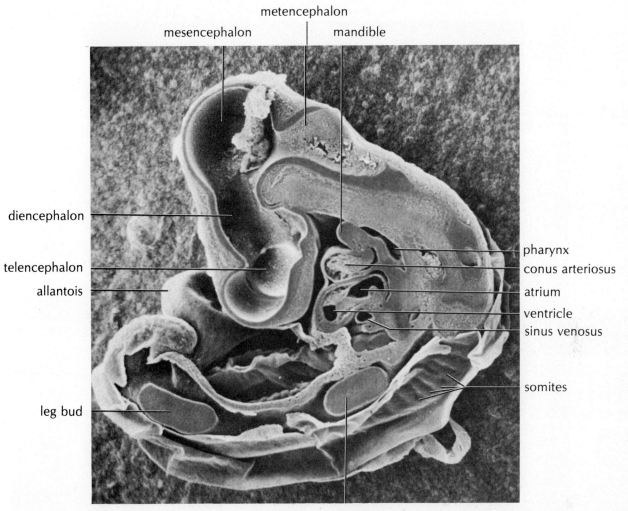

Figure 87. Chick embryo, 3–4 days, sagittal section, scanning electron micrograph. (27.6×)

mandible   conus arteriosus   pharynx

heart atrium

heart ventricle

sinus venosus

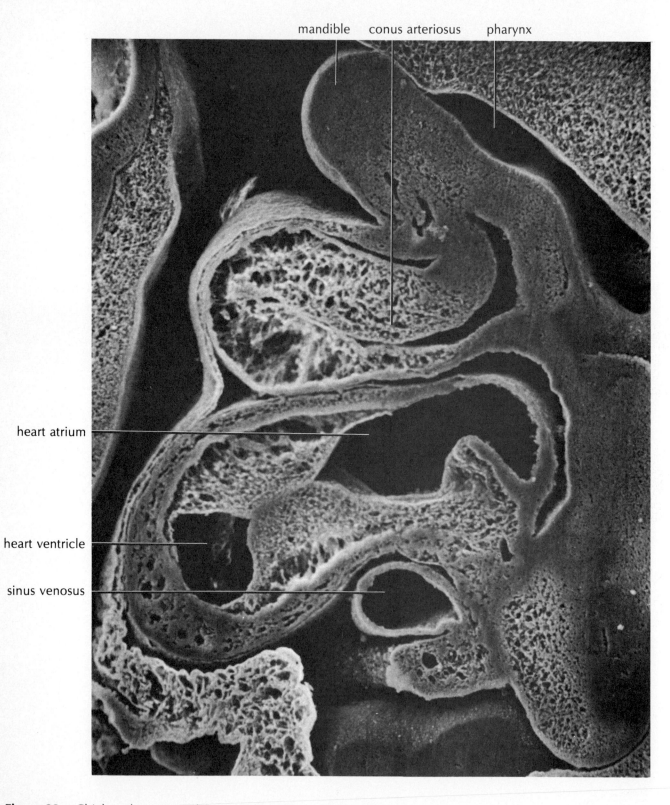

Figure 88. Chick embryo, 3–4 days, sagittal section of heart chambers, scanning electron micrograph. (141×)

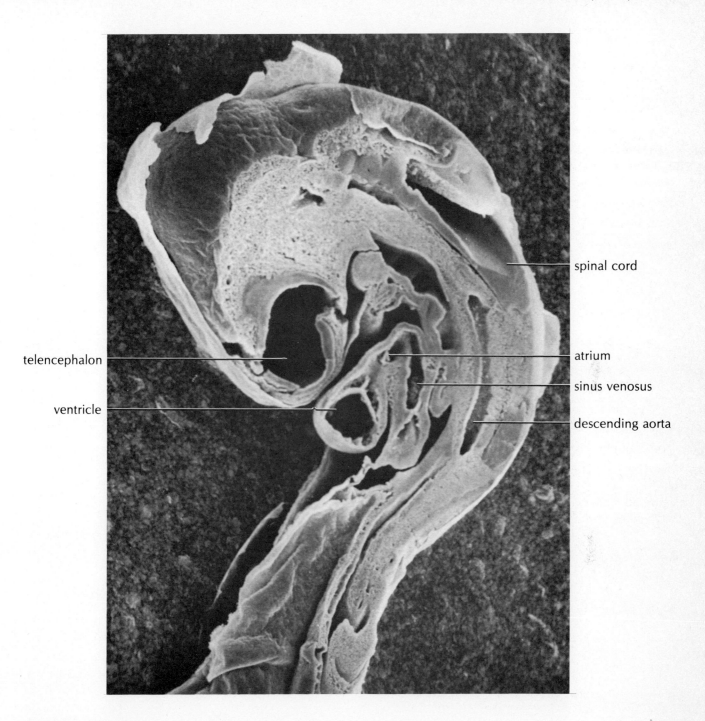

telencephalon

ventricle

spinal cord

atrium

sinus venosus

descending aorta

Figure 89.   Chick embryo, 3–4 days, sagittal section, scanning electron micrograph. (51.5×)

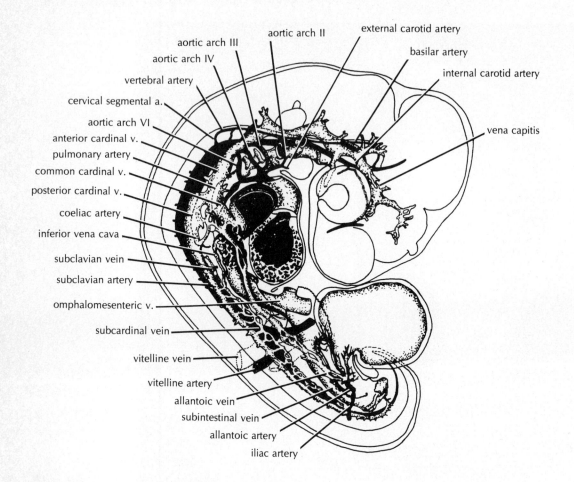

aortic arch III
aortic arch IV
vertebral artery
cervical segmental a.
aortic arch VI
anterior cardinal v.
pulmonary artery
common cardinal v.
posterior cardinal v.
coeliac artery
inferior vena cava
subclavian vein
subclavian artery
omphalomesenteric v.
subcardinal vein
vitelline vein
vitelline artery
allantoic vein
subintestinal vein
allantoic artery
iliac artery

aortic arch II
external carotid artery
basilar artery
internal carotid artery
vena capitis

Figure 90. Chick embryo, 4 days, circulatory system. From B. M. Patten, *Early Embryology of the Chick,* McGraw-Hill, 1957.

branchial clefts

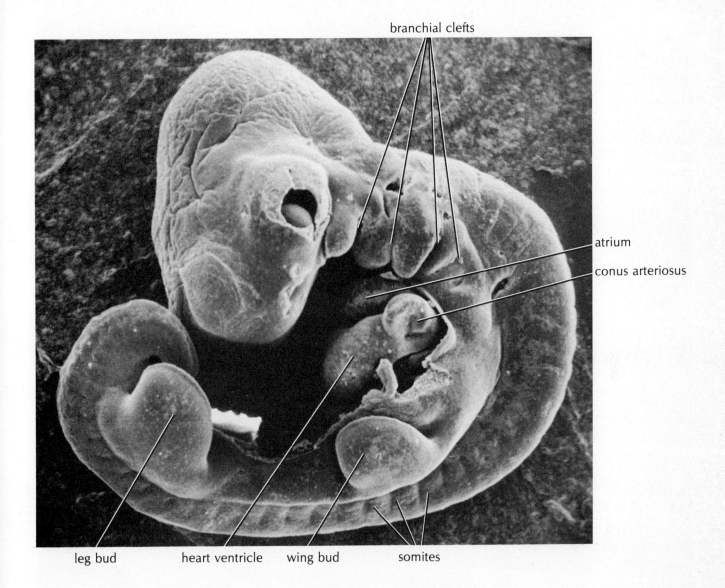

atrium

conus arteriosus

leg bud   heart ventricle   wing bud   somites

Figure 91. Chick embryo, 3–4 days, whole, scanning electron micrograph. (33.1×)

telencephalon   eye cup   lens vesicle   otic vesicle

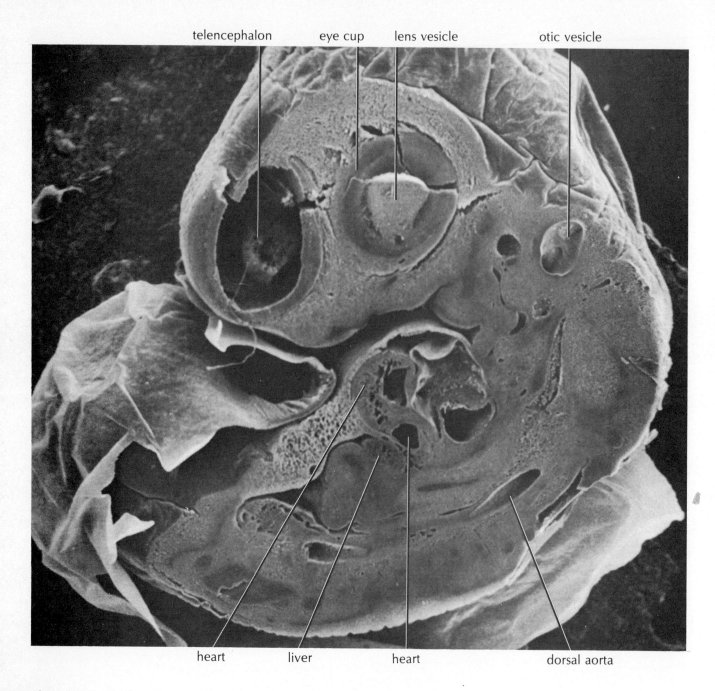

heart   liver   heart   dorsal aorta

Figure 92. Chick embryo, 3–4 days, sagittal section, scanning electron micrograph. (32.2×)

heart   anterior vitelline vein

vitelline arteries and
vitelline veins

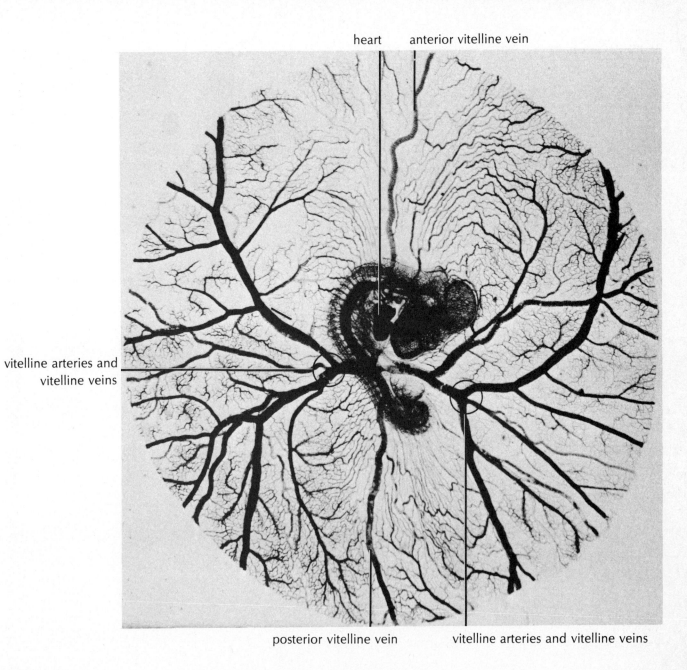

posterior vitelline vein   vitelline arteries and vitelline veins

Figure 93.   Chick embryo, 4 days, whole mount showing vitelline circulation.
(73×)

III
The Mammal

A. Cat and Rat Gonads

connective tissue   young Graafian follicle

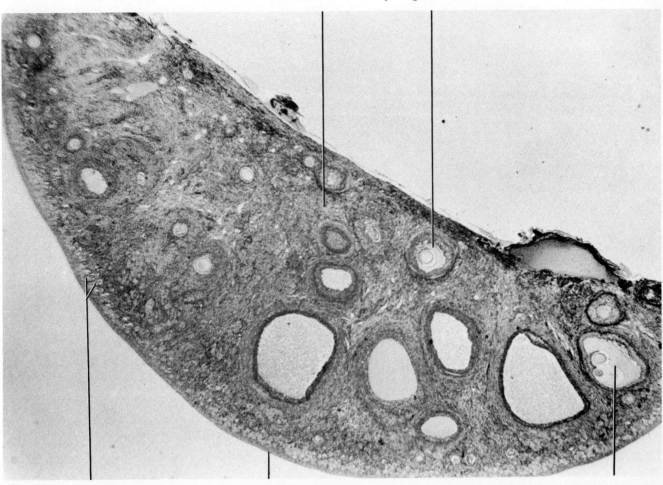

dormant primary follicles   tunica albuginea   Graafian follicle

Figure 94.   Cat ovary. (71×)

follicular cavity

theca interna

oocyte

cumulus oophorus

stratum granulosa

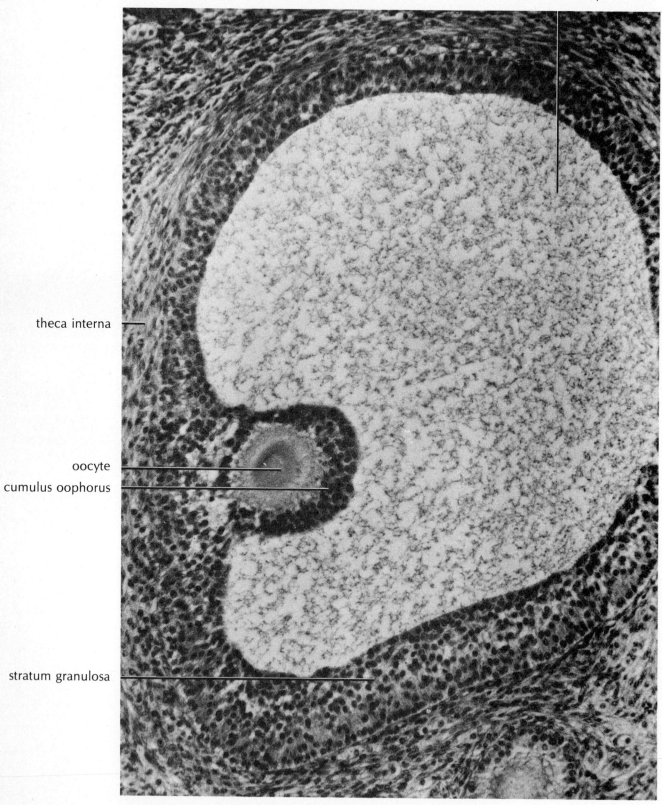

Figure 95. Graafian follicle of cat ovary. (268.7 ×)

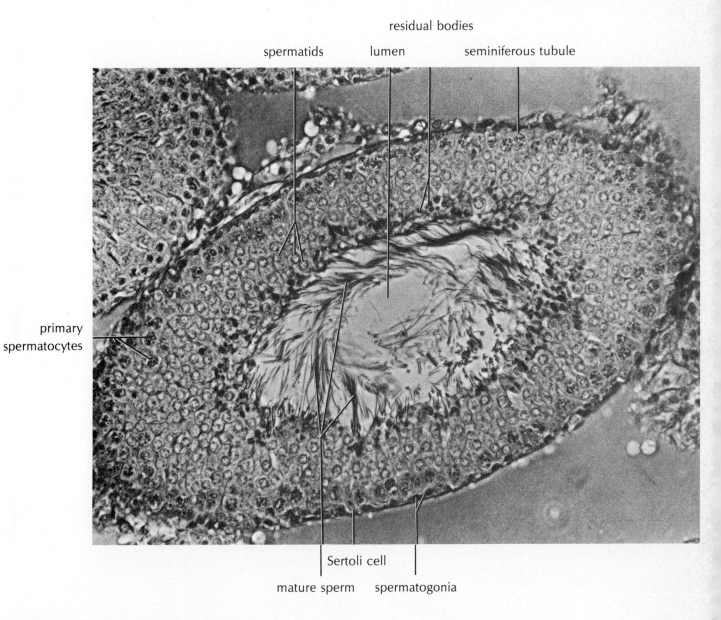

residual bodies

spermatids   lumen   seminiferous tubule

primary
spermatocytes

Sertoli cell

mature sperm   spermatogonia

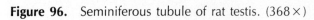

Figure 96. Seminiferous tubule of rat testis. (368×)

B. Pig, 6 mm Embryo

tongue

metencephalon   hyoid arch

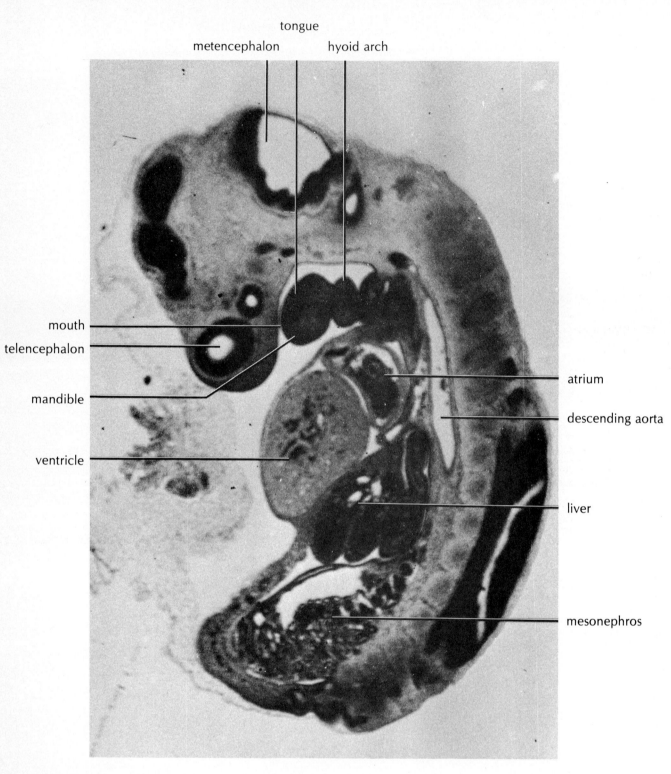

mouth

telencephalon

mandible

ventricle

atrium

descending aorta

liver

mesonephros

Figure 97. Pig embryo, 6 mm, sagittal section. (26.6x)

metencephalon   tongue   myelencephalon

mesencephalon

diencephalon
mandible

telencephalon

pericardial cavity

atrium

lung bud

ventricle

stomach

liver

descending aorta

spinal cord

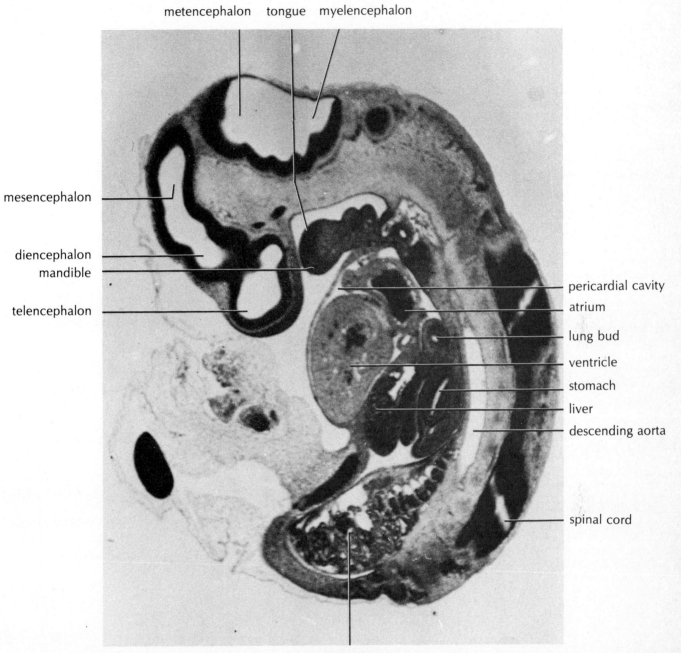

mesonephros

Figure 98.   Pig embryo, 6 mm, sagittal section. (23 ×)

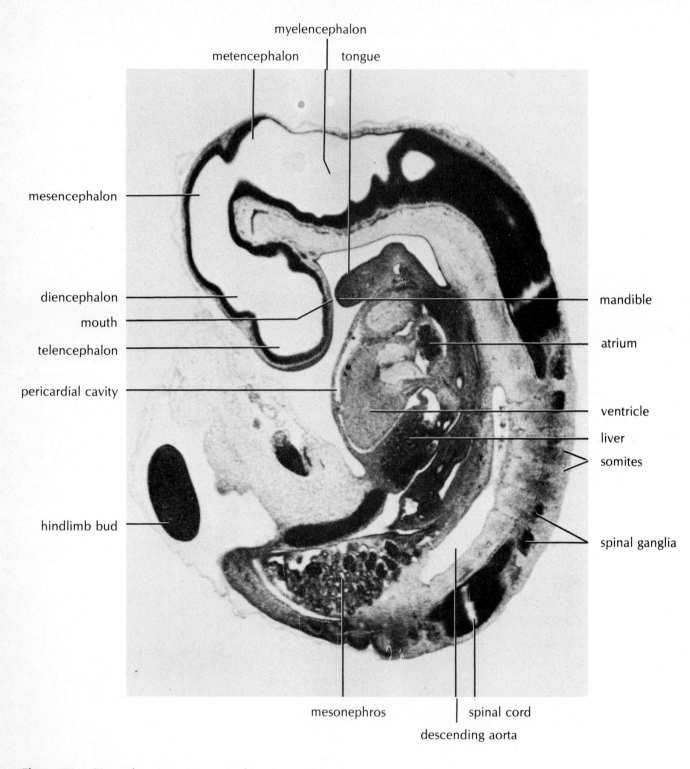

Figure 99. Pig embryo, 6 mm, sagittal section. (22.7×)

C. Pig, 10 mm Embryo

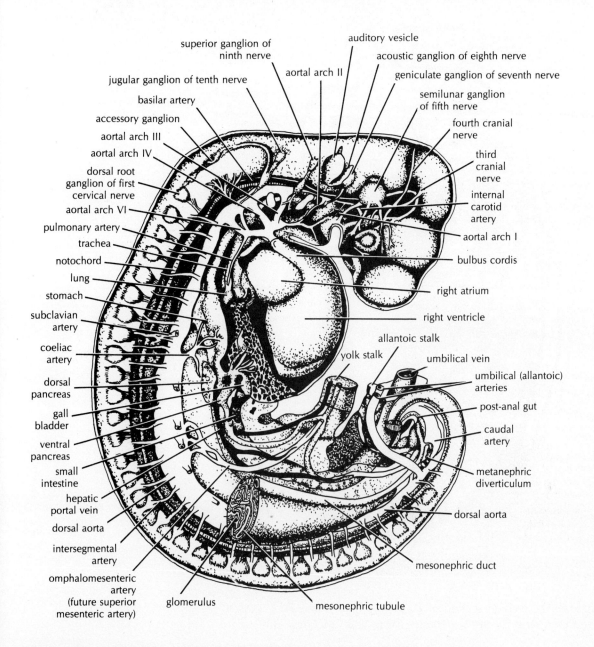

superior ganglion of
ninth nerve

jugular ganglion of tenth nerve

basilar artery

accessory ganglion

aortal arch III

aortal arch IV

dorsal root
ganglion of first
cervical nerve

aortal arch VI

pulmonary artery

trachea

notochord

lung

stomach

subclavian
artery

coeliac
artery

dorsal
pancreas

gall
bladder

ventral
pancreas

small
intestine

hepatic
portal vein

dorsal aorta

intersegmental
artery

omphalomesenteric
artery
(future superior
mesenteric artery)

auditory vesicle

aortal arch II

acoustic ganglion of eighth nerve

geniculate ganglion of seventh nerve

semilunar ganglion
of fifth nerve

fourth cranial
nerve

third
cranial
nerve

internal
carotid
artery

aortal arch I

bulbus cordis

right atrium

right ventricle

allantoic stalk

yolk stalk

umbilical vein

umbilical (allantoic)
arteries

post-anal gut

caudal
artery

metanephric
diverticulum

dorsal aorta

mesonephric duct

glomerulus

mesonephric tubule

Figure 100. Pig embryo, 10 mm, reconstruction. From O. E. Nelsen, *Comparative Embryology of the Vertebrates* (New York: McGraw-Hill, 1953).

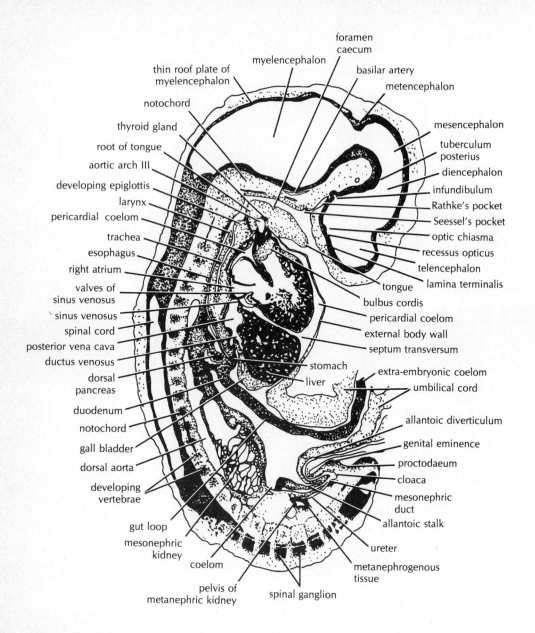

Figure 101. Pig embryo, median section of 10 mm embryo. From O. E. Nelsen, *Comparative Embryology of the Vertebrates* (New York: McGraw-Hill, 1953).

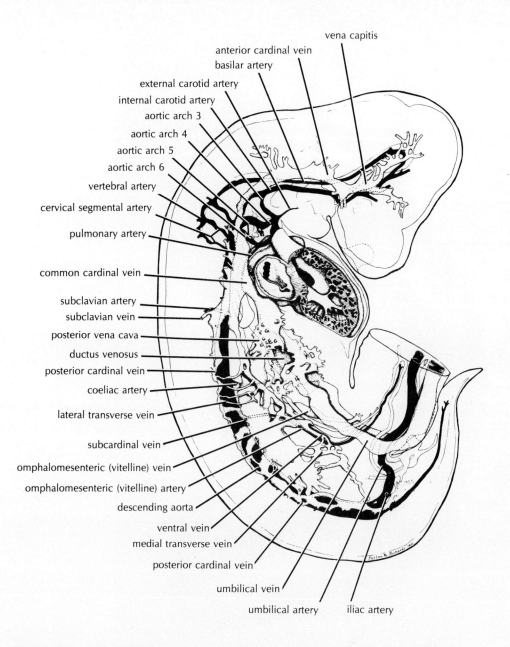

vena capitis
anterior cardinal vein
basilar artery
external carotid artery
internal carotid artery
aortic arch 3
aortic arch 4
aortic arch 5
aortic arch 6
vertebral artery
cervical segmental artery
pulmonary artery
common cardinal vein
subclavian artery
subclavian vein
posterior vena cava
ductus venosus
posterior cardinal vein
coeliac artery
lateral transverse vein
subcardinal vein
omphalomesenteric (vitelline) vein
omphalomesenteric (vitelline) artery
descending aorta
ventral vein
medial transverse vein
posterior cardinal vein
umbilical vein
umbilical artery   iliac artery

Figure 102.   Reconstruction of the circulatory system of a 9.4 mm pig embryo. From B. M. Patten, *Embryology of the Pig,* third edition (New York: McGraw-Hill, 1959).

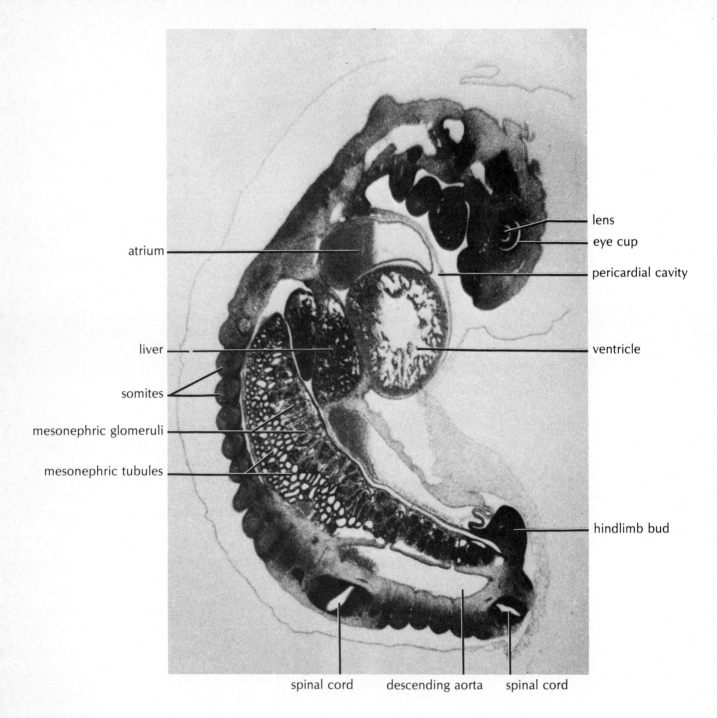

atrium

lens

eye cup

pericardial cavity

liver

ventricle

somites

mesonephric glomeruli

mesonephric tubules

hindlimb bud

spinal cord   descending aorta   spinal cord

Figure 103. Pig embryo, 10 mm, sagittal section. (22.2×)

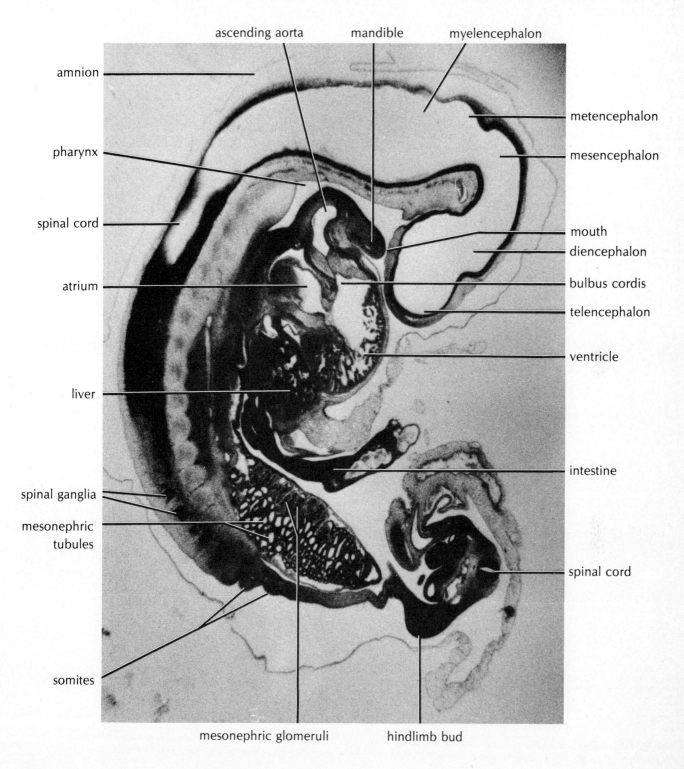

ascending aorta   mandible   myelencephalon

amnion

pharynx

spinal cord

metencephalon

mesencephalon

mouth

diencephalon

atrium

bulbus cordis

telencephalon

ventricle

liver

intestine

spinal ganglia

mesonephric
tubules

spinal cord

somites

mesonephric glomeruli   hindlimb bud

Figure 104.   Pig embryo, 10 mm, sagittal section. (21.5×)

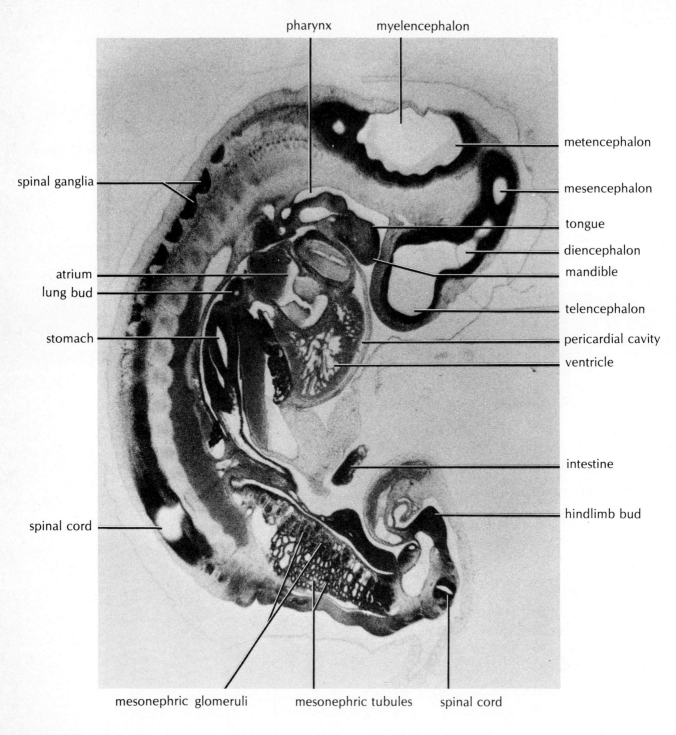

pharynx

myelencephalon

metencephalon

spinal ganglia

mesencephalon

tongue

diencephalon

atrium

mandible

lung bud

telencephalon

stomach

pericardial cavity

ventricle

intestine

spinal cord

hindlimb bud

mesonephric glomeruli   mesonephric tubules   spinal cord

Figure 105. Pig embryo, 10 mm, sagittal section. (20.9×)

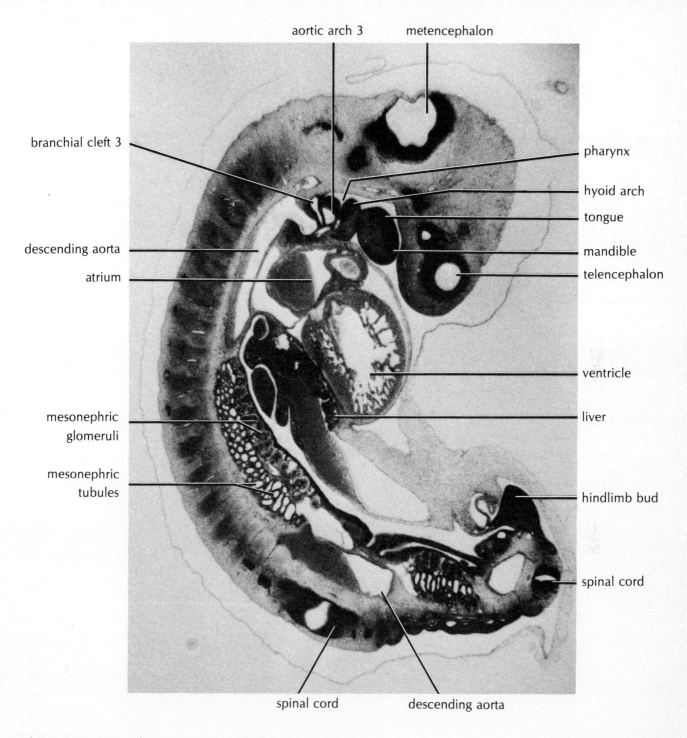

Figure 106. Pig embryo, 10 mm, sagittal section. (22.2×)

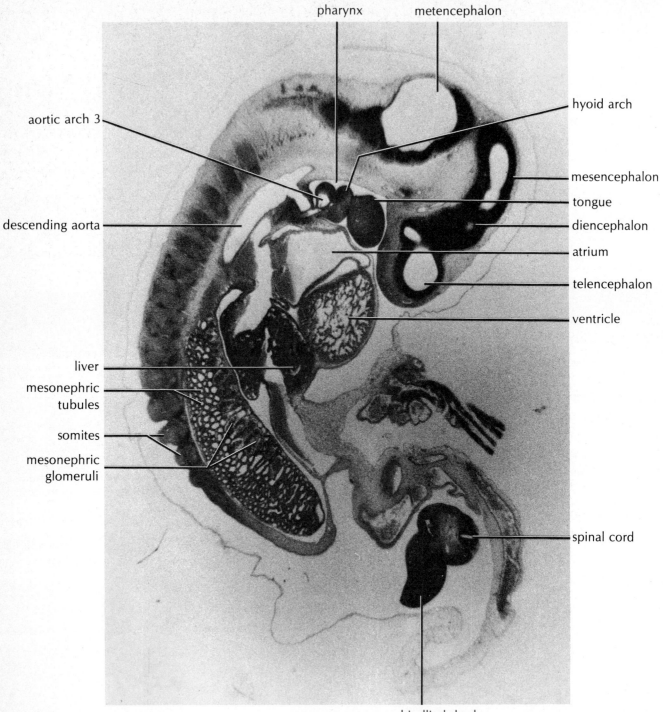

Figure 107.   Pig embryo, 10 mm, sagittal section. (20.6×)

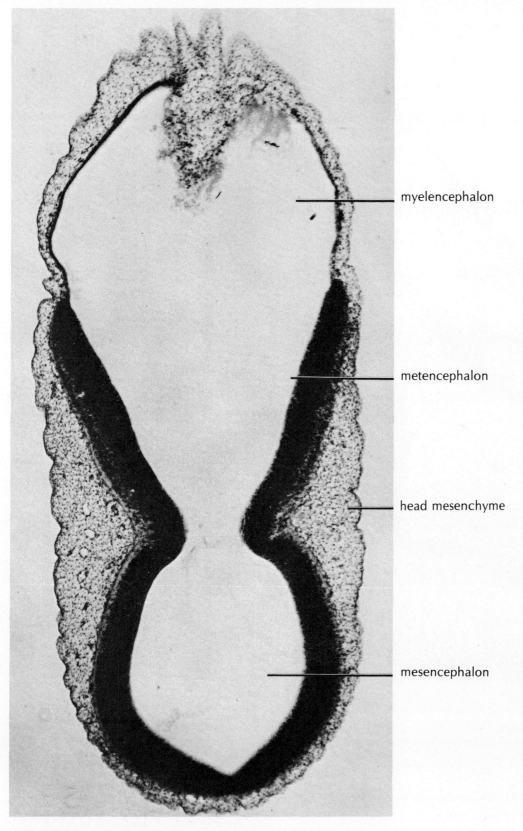

myelencephalon

metencephalon

head mesenchyme

mesencephalon

Figure 108. Pig embryo, 10 mm, transverse section through midbrain and hindbrain. (65×)

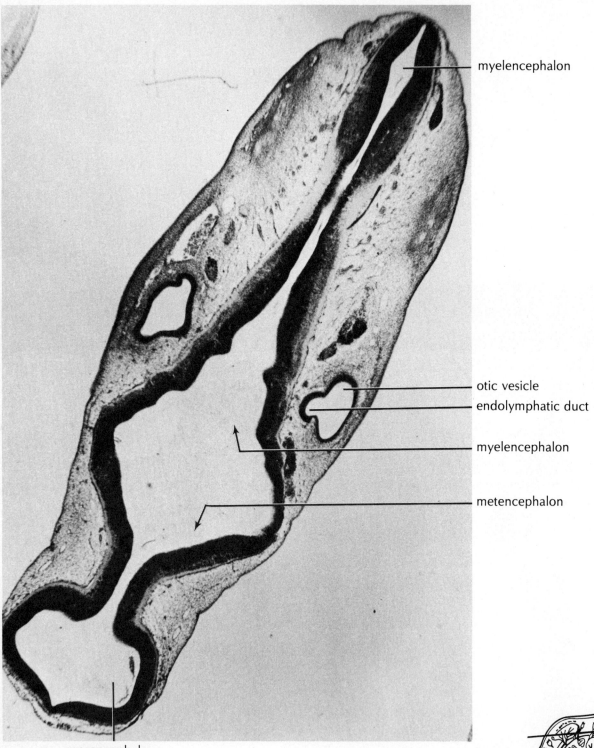

— myelencephalon

— otic vesicle
— endolymphatic duct

— myelencephalon

— metencephalon

mesencephalon

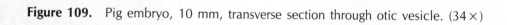

Figure 109. Pig embryo, 10 mm, transverse section through otic vesicle. (34×)

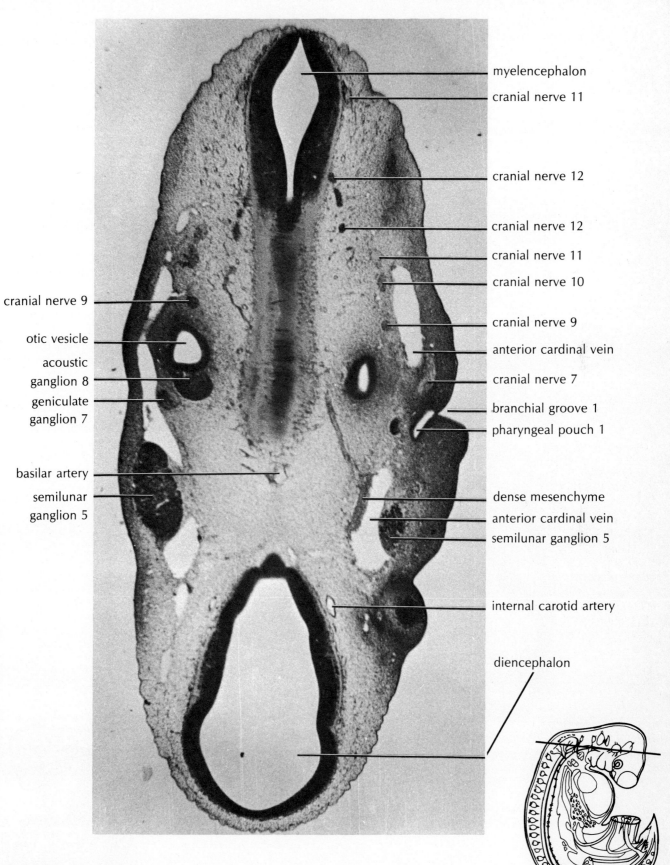

myelencephalon

cranial nerve 11

cranial nerve 12

cranial nerve 12

cranial nerve 11

cranial nerve 10

cranial nerve 9

anterior cardinal vein

cranial nerve 7

branchial groove 1

pharyngeal pouch 1

dense mesenchyme

anterior cardinal vein

semilunar ganglion 5

internal carotid artery

diencephalon

cranial nerve 9

otic vesicle

acoustic
ganglion 8

geniculate
ganglion 7

basilar artery

semilunar
ganglion 5

Figure 110. Pig embryo, 10 mm, transverse section through diencephalon and myelencephalon. (37.7×)

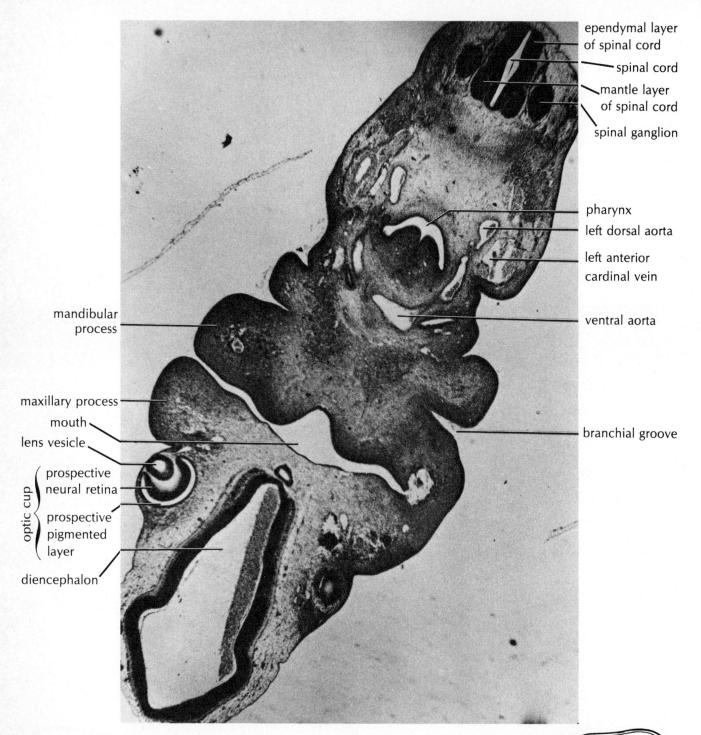

ependymal layer
of spinal cord

spinal cord

mantle layer
of spinal cord

spinal ganglion

pharynx

left dorsal aorta

left anterior
cardinal vein

ventral aorta

branchial groove

mandibular
process

maxillary process

mouth

lens vesicle

optic cup { prospective
neural retina

prospective
pigmented
layer

diencephalon

Figure 111.   Pig embryo, 10 mm, transverse section through a developing eye.
(32.8×)

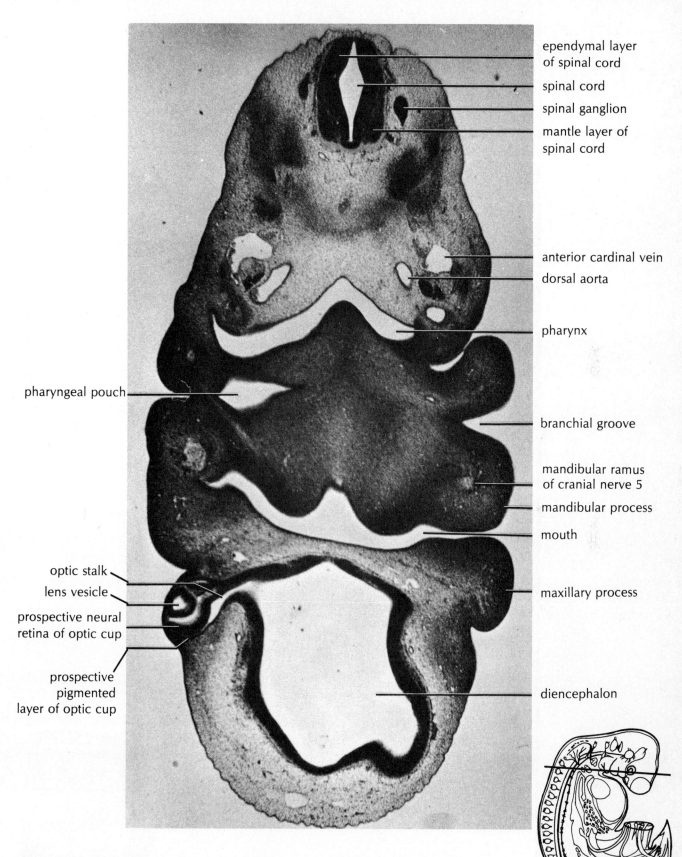

ependymal layer
of spinal cord

spinal cord

spinal ganglion

mantle layer of
spinal cord

anterior cardinal vein

dorsal aorta

pharynx

pharyngeal pouch

branchial groove

mandibular ramus
of cranial nerve 5

mandibular process

mouth

optic stalk

lens vesicle

prospective neural
retina of optic cup

maxillary process

prospective
pigmented
layer of optic cup

diencephalon

Figure 112. Pig embryo, 10 mm, transverse section through maxillary and mandibular processes. (37.7×)

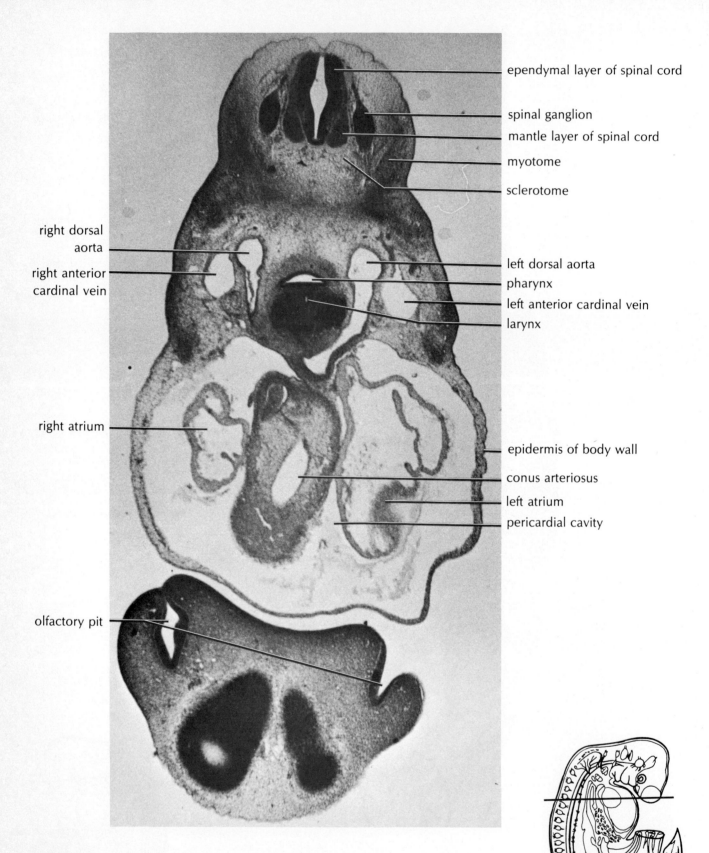

ependymal layer of spinal cord

spinal ganglion

mantle layer of spinal cord

myotome

sclerotome

right dorsal aorta

right anterior cardinal vein

left dorsal aorta

pharynx

left anterior cardinal vein

larynx

right atrium

epidermis of body wall

conus arteriosus

left atrium

pericardial cavity

olfactory pit

Figure 113. Pig embryo, 10 mm, transverse section through heart chambers. (38.5×)

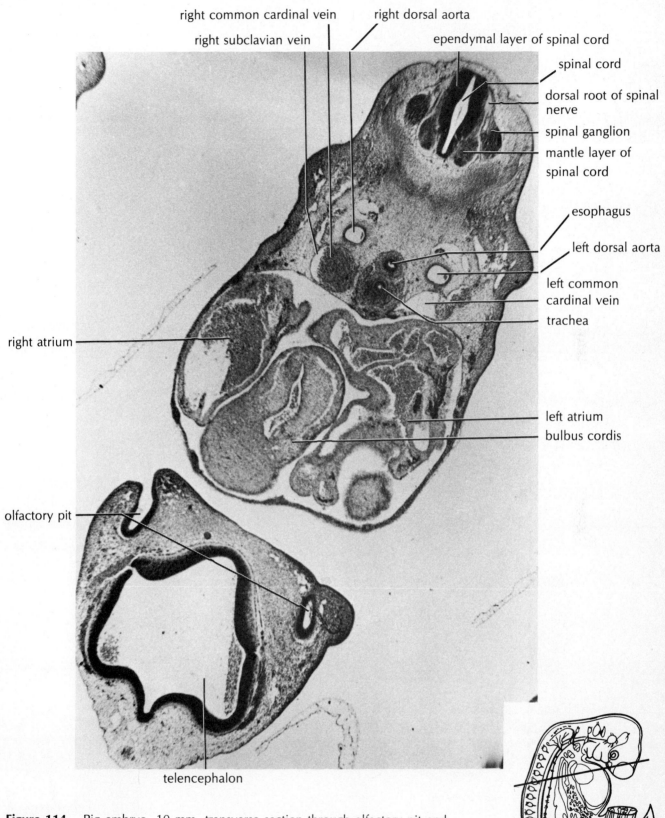

right common cardinal vein

right subclavian vein

right dorsal aorta

ependymal layer of spinal cord

spinal cord

dorsal root of spinal nerve

spinal ganglion

mantle layer of spinal cord

esophagus

left dorsal aorta

left common cardinal vein

trachea

right atrium

left atrium

bulbus cordis

olfactory pit

telencephalon

Figure 114. Pig embryo, 10 mm, transverse section through olfactory pit and heart chambers. (33.2×)

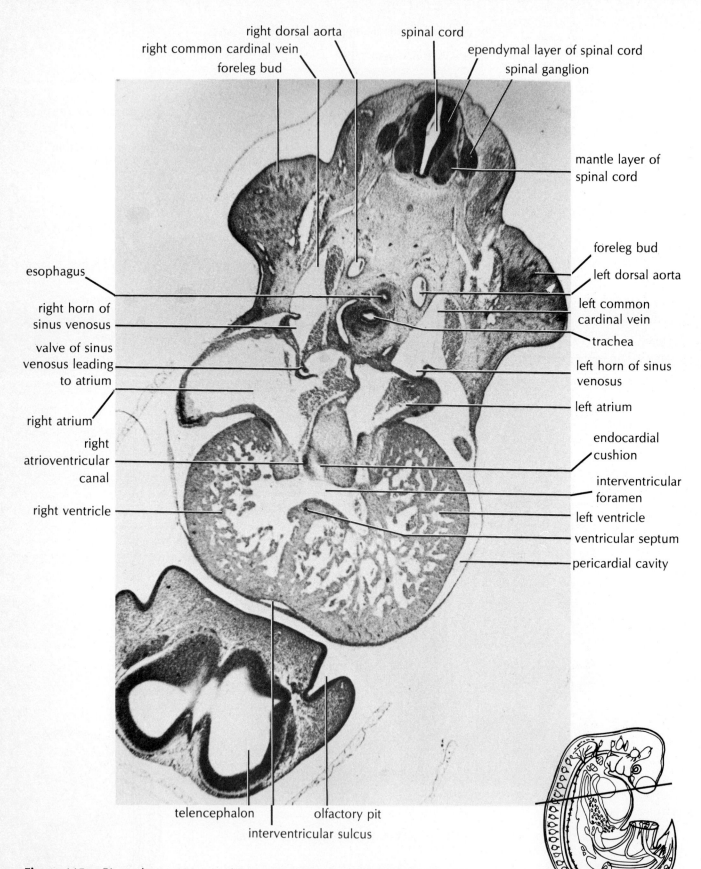

right dorsal aorta

right common cardinal vein

foreleg bud

spinal cord

ependymal layer of spinal cord

spinal ganglion

mantle layer of spinal cord

foreleg bud

left dorsal aorta

left common cardinal vein

trachea

esophagus

right horn of sinus venosus

valve of sinus venosus leading to atrium

right atrium

right atrioventricular canal

right ventricle

left horn of sinus venosus

left atrium

endocardial cushion

interventricular foramen

left ventricle

ventricular septum

pericardial cavity

telencephalon

interventricular sulcus

olfactory pit

Figure 115. Pig embryo, 10 mm, transverse section through foreleg buds and heart ventricles. (33.2×)

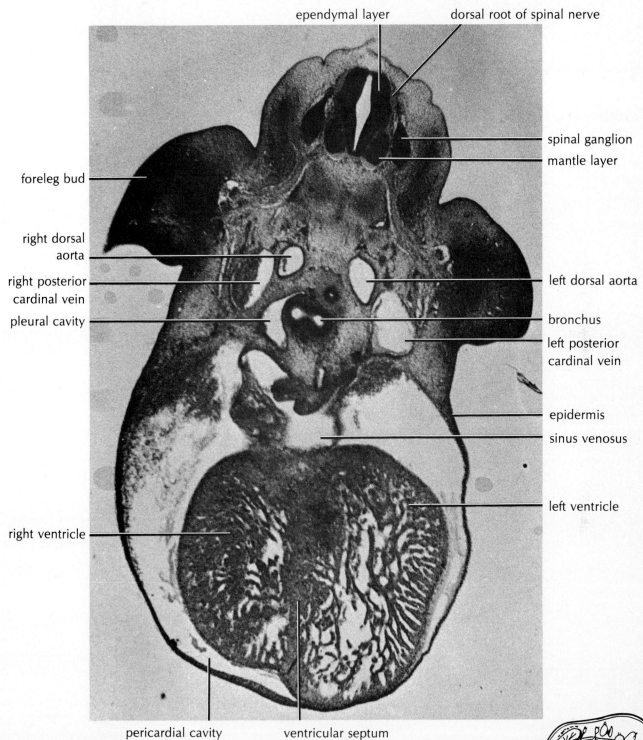

ependymal layer

dorsal root of spinal nerve

spinal ganglion

mantle layer

foreleg bud

right dorsal aorta

left dorsal aorta

right posterior cardinal vein

pleural cavity

bronchus

left posterior cardinal vein

epidermis

sinus venosus

left ventricle

right ventricle

pericardial cavity

ventricular septum

Figure 116.   Pig embryo, 10 mm, transverse section through heart ventricles and bronchi. (37.7×)

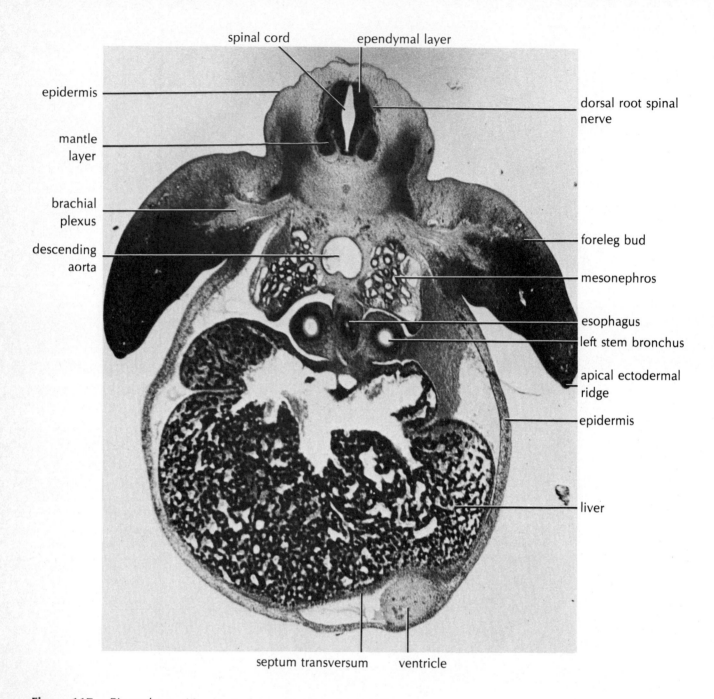

spinal cord   ependymal layer

epidermis

mantle
layer

brachial
plexus

descending
aorta

dorsal root spinal
nerve

foreleg bud

mesonephros

esophagus

left stem bronchus

apical ectodermal
ridge

epidermis

liver

septum transversum   ventricle

Figure 117.   Pig embryo, 10 mm, transverse section through liver. (34×)

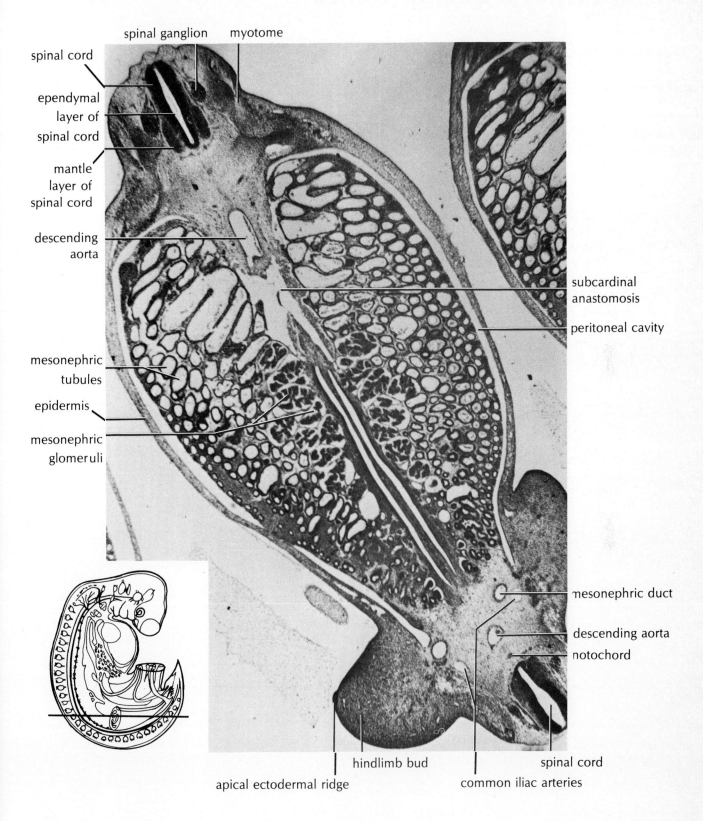

spinal ganglion

myotome

spinal cord

ependymal layer of spinal cord

mantle layer of spinal cord

descending aorta

subcardinal anastomosis

peritoneal cavity

mesonephric tubules

epidermis

mesonephric glomeruli

mesonephric duct

descending aorta

notochord

hindlimb bud

spinal cord

apical ectodermal ridge

common iliac arteries

Figure 118. Pig embryo, 10 mm, transverse section through mesonephros. (35.2×)

D.   Human Development

Table 3.   Some Major Characteristics of Human Development

Week	Characteristics
1	Fertilization, cleavage, morula, blastocyst, implantation begins, endoderm of embryo visible
2	Embryonic disc is bilaminar, amniotic cavity appears, implantation complete, primitive placental circulation established
3	Primitive streak forms, gastrulation occurs, neurulation begins
4	Heart begins to beat, neural folds fusing, eye and ear primordia present, 4 pairs of branchial arches, arm and leg buds present
5	Lens vesicles, optic cups, nasal pits form; hand and foot plates are paddle shaped; heart atrium dividing
6	Oral and nasal cavities confluent, upper lip formed, arms bent at elbow, fingers distinct but webbed, palate developing
7	Eyelids forming, tip of nose distinct, genital tubercle, urogenital and anal membranes form, trunk elongating and straightening
8	Upper limbs bent at elbows, fingers distinct, anal membrane perforated; urogenital membrane degenerating, testes and ovaries distinguishable, external genitalia indifferent but beginning to differentiate, all essential external and internal structures have begun to form
9	Begins fetal period, genitalia differentiation begins
10	Face has human appearance, genitalia begin to show distinct male or female characteristics
11	Growth and elaboration of all structures continues; at about 12 weeks the sex of the fetus is externally distinguishable

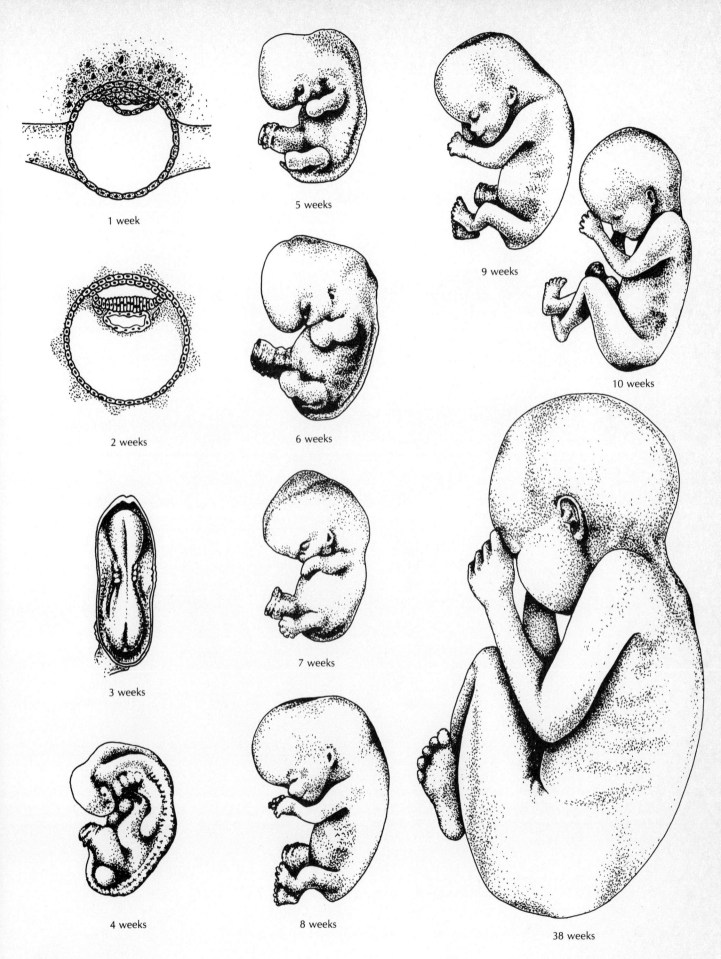

1 week

2 weeks

3 weeks

4 weeks

5 weeks

6 weeks

7 weeks

8 weeks

9 weeks

10 weeks

38 weeks

Figure 119. Summary of human development.

133

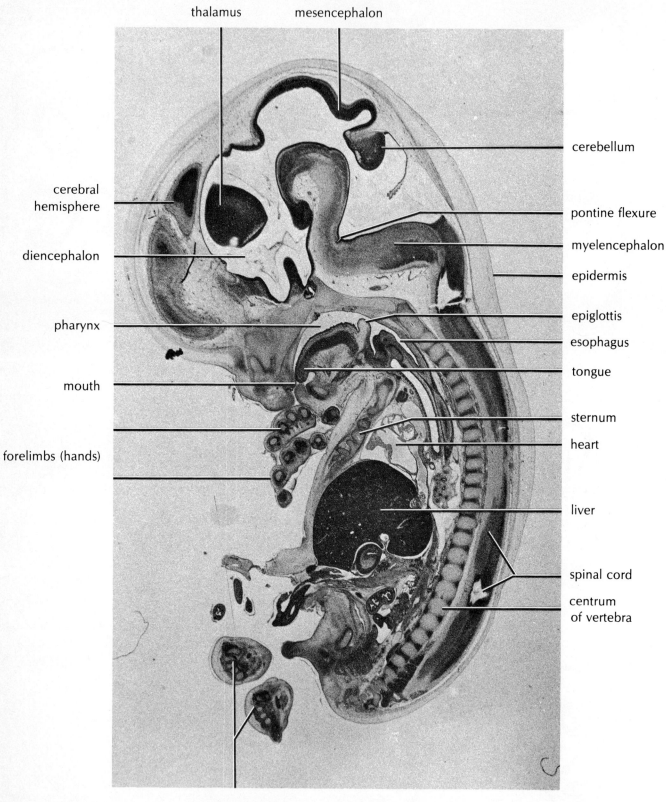

thalamus   mesencephalon

cerebellum

cerebral
hemisphere

pontine flexure

myelencephalon

diencephalon

epidermis

pharynx

epiglottis

esophagus

mouth

tongue

sternum

heart

forelimbs (hands)

liver

spinal cord

centrum
of vertebra

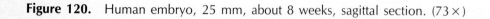

hindlimbs (feet)

Figure 120.   Human embryo, 25 mm, about 8 weeks, sagittal section. (73×)

Appendix

Planes and Sections Terminology
Used in this Atlas

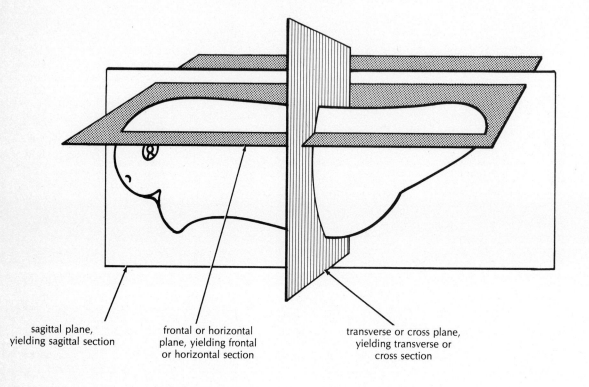

sagittal plane,
yielding sagittal section

frontal or horizontal
plane, yielding frontal
or horizontal section

transverse or cross plane,
yielding transverse or
cross section

Obtaining Frog Embryos for Microscopy for this Atlas

Male and female leopard frogs (*Rana pipiens*) were obtained from commercial dealers. Frogs were stored at 4°C until ready for use, at which time they were transferred to room temperature. Female frogs in prebreeding condition were inoculated with frog pituitary gland extract obtained from Carolina Biological Supply Company to induce ovulation. This extract, made from dried pituitary glands, was mixed with distilled water and then injected into the abdominal cavities of the female frogs. Eggs were released (*stripped*) from the female frogs and fertilized by gently squeezing the frogs over a sperm suspension prepared by mincing frog testes in spring water or 10% Holtfreter's solution or 10% Amphibian Ringer's solution (formulae given in this appendix). The fertilized frog eggs were maintained in spring water or one of the above mentioned 10% solutions at room temperature. Development proceeded according to the timetable given earlier in the text. At selected times, embryos were placed in buffer and fixative and prepared for microscopy.

Obtaining Chick Embryos for Microscopy for this Atlas

Chick eggs at various stages of development were cracked into finger bowls containing Howard Ringer's solution (formula in this appendix). The blastoderm, or chick embryo proper, is seen as a whitish colored disc that lies atop the yolk. The blastoderm was removed from the yolk by cutting around the entire blastoderm with a pair of scissors and gently peeling it off the yolk into the Ringer's solution. The vitelline membrane was usually removed from the blastoderm if it was still attached after the blastoderm was removed from the yolk by gently teasing it loose around the edges of the blastoderm and lifting it away. The blastoderm was then placed into buffer and fixative by sucking it out of the Ringer's solution with a wide-mouth pipette.

Preparation of Specimens for Scanning Electron Microscopy for this Atlas

The embryos used for the scanning electron micrographs were placed in 0.1M sodium cacodylate buffer solution (pH 7.2–7.3) and fixed in 2.5% glutaraldehyde, 4% paraformaldehyde and 0.075% $CaCl_2$ in 0.1M sodium cacodylate buffer for 8–24 hours at 4°C. They were then washed overnight at 4°C in 0.1M sodium cacodylate buffer and postfixed in 1% osmium tetroxide in 0.1M sodium cacodylate buffer. Embryos were dehydrated through a graded ethanol series of 50%, 70%, 80%, 95%, 100% ethanol. Eosin (0.05%) in 95% ethanol was used to stain the embryos for identification. The embryos were cleared in toluene, oriented, and embedded in paraffin (56–57°C). The embryos were sectioned with an American Optical AO 820 Microtome, and sections were examined by magnifier and/or dissecting microscope to determine if the desired depth of sectioning was achieved. Paraffin was removed from the remainder of the embryo in several changes of toluene at 57°C and embryos were transferred through 100% ethanol twice. Embryos were critical point dried in liquid CO_2. Dried embryos were mounted on scanning electron microscopy stubs with silver conductive paint and coated with gold (200 Å thick coating) with a Technics Hummer II sputtering coating device. Embryos were examined and photographed with an International Scientific Instruments Super II Scanning Electron Microscope operated at 15 KV.

References Used for Preparing Embryos for Scanning Electron Microscopy

Armstrong, P.B., A scanning electron microscope technique for study of the internal microanatomy of embryos, *Microscope* 19:281–284 (1971).

Armstrong, P.B. and D. Parenti, Scanning electron microscopy of the chick embryo, *Developmental Biology* 33:457–462 (1973).

Karnovsky, M.J., A formaldehyde-glutaraldehyde fixative of high osmolality for use in electron microscopy, *Journal of Cell Biology* 27:137A–138A (1965).

Light Microscopy Equipment Used for this Atlas

Sections larger than 4mm in diameter were studied and photographed with a Wild M-5 stereomicroscope and a Nikon F2A camera. All other sections were studied and photographed using a Zeiss Photomicroscrope I. Panatomic-X (ASA 32) 35 mm film was used for all micrographs. We thank Professors Anthony Gaudin, Daisy Kuhn, and George Lefevre for permitting us to use their equipment.

Holtfreter's Solution

An amphibian Ringer's solution that is usually used at 10% strength for culturing amphibian material.

Full Strength Formula

Component	Quantity
NaCl	3.5 gm
KCl	0.05 gm
$CaCl_2$	0.1 gm
$NaHCO_3$	0.2 gm
Distilled water	1000 ml

Amphibian Ringer's Solution

An amphibian Ringer's solution that is usually used at 10% strength for culturing amphibian material.

Full Strength Formula

Component	Quantity
NaCl	6.60 gm
KCl	0.15 gm
$CaCl_2$	0.15 gm
$NaHCO_3$	enough to adjust pH to 7.8
Distilled water	1000 ml

Howard Ringer's Solution

This is a good isotonic saline solution for use with living chick material.

Component	Quantity
NaCl	7.2 gm
CaCl$_2$ (2H$_2$O)	0.17 gm (or 0.23 gm)
KCl	0.37 gm
Distilled water	1000 ml

Glossary

acoustic ganglion 8: Ganglion of cranial nerve 8.

adhesive gland: Ectodermal thickening on the ventral side of the head of frog tadpoles used for attachment to a substratum (such as rocks, plants, etc.)

allantois: Extraembryonic saclike extension of the hindgut of amniotes, serving excretion and respiration.

amnion: Extraembryonic membrane of amniotes, inside the chorion, composed of somatopleure.

amniotes: Vertebrates possessing an amnion during development.

amniotic cavity: Space between amnion and embryo proper.

animal region (or animal hemisphere): Region of egg where the nucleus resides, opposite the vegetal region.

anterior cardinal veins: Primitive paired veins of the head.

anterior intestinal portal: Opening from midgut into foregut in amniotes.

anterior neuropore: Temporary opening into neural tube.

anus: Posterior opening of the digestive tube.

aorta: Main trunk of the arterial system.

aortic arches: Paired arterial connections between the dorsal and ventral aortae.

apical ectodermal ridge: Ectodermal thickening on limb bud tip.

area opaca: Peripheral zone of the chick blastoderm which is attached to the yolk below.

area pellucida: Relatively transparent central region of the chick blastoderm underlaid by the subgerminal space.

archenteron: Primitive embryonic digestive tube.

archenteron roof: The dorsal covering of the archenteron that becomes the notochord.

ascending aorta: Portion of aorta that extends anteriorly from the heart.

atrioventricular canal: Passage connecting the heart atrium and ventricle.

atrium (of heart): Heart chamber that delivers blood to the ventricle. In the mammalian heart, the right atrium receives venous blood from the body and delivers it to the right ventricle. The left atrium receives oxygenated blood from the lungs via the pulmonary vein and delivers it to the left ventricle.

basilar artery: Median artery below the hindbrain connecting the vertebral arteries and the internal carotid arteries.

blastocoel: Cavity of bastula stage embryos.

blastoderm: The primitive cellular plate of early embryos.

blastopore: Opening into the archenteron from outside the embryo. This term is used in early embryos; later on, the blastopore region differentiates into other structures such as the anus.

blastula: The embryonic stage in which the embryo is a hollow ball with a cavity, or is a cap of cells above a cavity.

brachial plexus: Interconnection of thoracic and spinal nerves from which the forelimb nerves branch.

branchial arches: Series of paired bars in the wall of the pharynx, that give rise to structures including parts of the jaws, skull, and middle ear.

branchial clefts: Series of paired perforations in the wall of the pharynx that separate the branchial arches from each other.

branchial grooves: Paired ectodermal invaginations in the wall of the pharynx. Each groove corresponds to a branchial (pharyngeal) pouch.

branchial pouches: See pharyngeal pouches. These are endodermal evaginations of the lateral wall of the pharynx that form parts of the middle ear, tonsil, thymus, and parathyroid glands.

bronchus: Respiratory tube connecting the trachea with the lungs.

bulbus arteriosus: See *conus arteriosus*.

bulbus cordis: See *conus arteriosus*.

chorion: Outermost extraembryonic membrane of amniotes, composed of somatopleure; also used to describe a surface coat exterior to the plasma membrane in the eggs of fishes and tunicates.

coelom: Body cavity.

common cardinal vein: Trunk of the anterior and posterior cardinal veins connecting the sinus venosus.

common iliac arteries: Large terminal branches of the aorta.

conus arteriosus: (bulbus arteriosus, bulbus cordis) Anterior most portion of the heart, connecting the ventricle with the ventral aorta.

cornea: Transparent front covering of the eye.

cranial nerves (c.n.):   nerve pairs arising from the brain.

c.n. 3 innervate all inner eye muscles and some extrinsic eye muscles (also called *oculomotor* nerve).

c.n. 4 innervate superior oblique ocular muscles (also called *trochlear* nerve).

c.n. 5 innervate mandibular arch region (also called *trigeminal* nerve).

c.n. 6 innervate external rectus eye muscles.

c.n. 7 innervate the hyoid arch.

c.n. 8 innervate the inner ear (also called the *vestibulocochlear* nerve).

c.n. 9 innervate the third branchial arch (also called *glossopharyngeal* nerve).

c.n. 10 innervate branchial arches 4, 5, 6. In the frog tadpole innervate the lateral line (which forms sense organs in the epidermis) (also called *vagus* nerve).

c.n. 11 innervate muscles of shoulder and pharynx (also called *accessory* nerve).

c.n. 12 innervate tongue muscles (also called *hypoglossal* nerve).

cumulus oophorus: Follicle cell layers of stratum granulosa surrounding the oocytes of mammals.

dense mesenchyme: groups of mesenchyme cells (embryonic connective tissue) that will form cartilage.

dermatome: Outer region of the somite that gives rise to the dorsal dermis of the skin.

descending aorta: Main trunk artery formed by fusion of the paired dorsal aortae.

diencephalon: Posterior portion of the forebrain, that forms the thirst center (thalamus), hunger center (hypothalamus), the posterior pituitary gland, and the optic vesicles.

dormant primary follicles: Small capsules inside the outer ovary wall of mammals, each containing an immature oocyte and follicle cells.

dorsal aortae: Paired arteries which fuse together posterior to the pharynx to form the descending aorta.

dorsal lip of the blastopore: prospective notochord; the first region to enter the amphibian embryo during gastrulation.

dorsal mesocardium: Dorsal mesentery (supporting membrane) of the heart.

dorsal root of spinal nerve: Dorsal portion of spinal nerve connecting the nerve trunk with the alar plate (dorsal lateral wall) of the spinal cord.

ectoderm: Outer germ layer (original part) of the embryo.

endocardial cushion: Connective tissue ring that forms the valves in the canal between the heart atrium and ventricle.

endocardium: Inner lining of the heart.

endoderm: Innermost germ layer of the embryo.

endolymphatic duct: Stalk of the otic vesicle.

ependymal layer: Inner layer of primitive neuroepithelial cells (inner surface nerve cells) of the neural tube.

epidermis: Outer epithelial (surface) portion of the skin.

epimere: somite; dorsal region of mesoderm consisting of myotome, dermatome, and sclerotome.

epimyocardium: Outer layer of heart.

esophagus: Digestive tube connecting pharynx with stomach.

extraembryonic coelom: Cavity outside the embryo proper that is continuous with the body cavity (coelom) and is surrounded by extraembryonic membranes.

fertilization membrane: a coat that forms at the surface of eggs during fertilization.

follicle cells: Cells surrounding oocytes that help nourish and protect the developing oocytes.

follicular cavity (antrum): Space inside of the Graafian follicle filled with fluid.

forebrain: Anterior most portion of the brain.

foregut: Anterior portion of the gut tube, which connects with the midgut and forms the pharynx (and its derivatives), stomach, and duodenum.

ganglion: A group of nerve cells whose cell bodies are located outside the central nervous system.

geniculate ganglion 7: Ganglion of the seventh cranial nerve.

genital ridge: Thickening of posterior mesoderm that will form the gonad.

germinal vesicle: Enlarged nucleus of the oocyte.

Graafian follicle: Capsule in the mammalian ovary, containing a follicular cavity, an oocyte, and follicle cells.

Hensen's node (primitive knot): Thickened anterior end of the primitive streak.

hindbrain: Posterior portion of the brain that forms the cerebellum and medulla.

hindgut: Posterior portion of the digestive tube that forms the posterior part of the large intestine and the cloaca.

hyoid arch: The second branchial arch that forms parts of the cartilages in the head region.

hypomere (lateral plate mesoderm): The most ventral subdivision of the mesoderm, consisting of somatic and splanchnic mesoderm.

inferior vena cava: Main trunk vein.

infundibulum: A ventral evagination of the prosencephalon.

internal carotid arteries: Extensions of the dorsal aortae that provide the main arterial blood supply to the brain.

intersomitic grooves: Spaces separating somites.

interventricular foramen: Opening between the right and left ventricles of the heart.

interventricular sulcus: Groove on the surface of the heart ventricle marking the plane of its impending division into left and right ventricles.

intestine: Portion of the gut tube posterior to the stomach.

larynx: Voice box

leg bud: Rudiment of leg

lens: Light-focusing structure of the eye.

lens placode: Thickening of head epidermis that will form the lens of the eye.

lens vesicle: Sac resulting from an invagination of the lens placode which will form the eye lens.

limb bud: Rudiment of limb.

liver diverticulum: Evagination of gut that gives rise to the liver, gall bladder, and common bile duct.

lumen: Inner passage of a tubular structure.

macromeres: Large cleavage blastomeres.

mandible: Lower jaw.

mandibular arch: Branchial arch I.

mandibular process: Posterior division of mandibular arch.

mandibular ramus of cranial nerve 5: Posterior portion of cranial nerve 5, innervating the mandible and jaw muscles.

mantle layer: Middle layer of developing neural tube.

maxillary process: Anterior portion of the mandibular arch.

meiosis: Divisions of the germ cell line that eventually result in the formation of haploid gametes.

mesencephalon: Midbrain.

mesenchyme: Embryonic connective tissue.

mesoderm: Middle germ layer (original part) of the embryo.

mesomere: Intermediate mesoderm between epimere and hypomere; also used to describe medium-sized cleavage blastomeres.

mesonephric duct: Duct connecting mesonephric tubules and cloaca.

mesonephric glomeruli: Capillaries within Bowman's capsules (invaginated kidney tubules) of the mesonephros.

mesonephric tubules: Kidney tubules of adult fish and amphibians, and of embryonic birds and mammals.

metencephalon: Anterior portion of hindbrain, which gives rise to the cerebellum of the brain.

micromeres: Small cleavage blastomeres.

midgut: Area of digestive tube or prospective digestive tube between foregut and hindgut, which forms the jejunum and ileum (portions of the small intestine) and the anterior portion of the large intestine.

mouth: Anterior opening of the digestive tube.

myelencephalon: Posterior portion of hindbrain, which forms the medulla of the brain.

myotome: Somite division that forms the skeletal muscle of the body wall.

neural folds: Elevated ridges of the neural plate.

neural groove: Trough formed by the bending up of the neural plate.

neural plate: Embryonic region that becomes the nervous system.

neural retina: Sensory retina.

neural tube: Tube derived from the neural plate that forms the nervous system.

neuromeres: Constricted minor segments of the brain.

notochord: Fibrocellular rod constituting the primitive skeletal axis.

nuclear membrane: Membrane surrounding the nucleus.

nucleolus: Dense granule containing ribosomal RNA and proteins, found in the nucleus of cells. Site of ribosome assembly.

nucleus: Cellular organelle containing the chromosomes and usually one or more nucleoli.

olfactory pits: Cavities on the lateral region of head, arising by invagination of the olfactory placodes, that will form the nasal cavities.

olfactory placodes: Ectodermal thickenings on the lateral regions of the head that eventually form the nasal passages.

oocyte: An immature egg, developing from the oogonium, that through growth and meiosis gives rise to a mature egg.

oogonia: Primordial egg cells that give rise to oocytes and eggs.

optic cup: Invaginated outpocketing of the diencephalon of the brain that forms the neural retina and pigmented coat of the eye.

optic stalk: Connection of the optic cup to the diencephalon.

otic placode (ear placode; auditory placode):   Thickening of head epidermis that eventually forms the inner ear.

otic vesicle (ear vesicle; auditory vesicle):   Chamber formed from invagination of the otic placode which will form the inner ear.

pericardial coelom (pericardial cavity):   Cavity around the heart.

pharyngeal pouches: Paired evaginations of the lateral wall of the pharynx.

pharynx: Anterior portion of the foregut.

pigmented layer (of eye):   Outer wall of optic cup, forming a light-tight coat.

pleural cavity: Body cavity surrounding each lung.

posterior cardinal veins: Primitive paired veins of the trunk situated dorsal to the mesonephros (the functional kidney of adult fish and amphibians).

primary oocytes: Cells arising as a result of growth and DNA duplication in oogonia.

primary spermatocytes: Cells arising as a result of growth and DNA duplication in spermatogonia.

primitive knot: See Hensen's node.

primitive streak: Thickening in the surface of some embryos at the beginning of gastrulation.

proamnion: Cresent-shaped area around the head of early bird embryos.
prosencephalon: Forebrain.
prospective (presumptive) region: The region of an early embryo that will become a specific structure in the embryo's later development.

residual bodies: Particles of cytoplasm discarded by the spermatids as they differentiate into mature sperm.
rhombencephalon: Hindbrain.

sclerotome: Medial somite region that gives rise to the vertebral column.
secondary oocyte: A meiotic product of primary oocyte.
secondary spermatocyte: Product of the meiotic division of a primary spermatocyte.
segmental mesoderm: Prospective somites.
semilunar ganglion 5: Ganglion (aggregation of nerve cells) of cranial nerve 5.
seminiferous tubules: Sperm-forming tubules of the testis.
Sertoli cells: Cells in the testis that support and nourish developing sperm.
shrinkage artifact: Separation of cellular components or tissues due to the preparation of the material for study; absent in the living organism.
sinus venosus: Posterior chamber of the embryonic heart that receives venous blood.
somatic mesoderm: Hypomere mesoderm in close contact with ectoderm.
somatopleure: Combination of somatic mesoderm and ectoderm.
somite: Epimere; dorsal region of mesoderm consisting of myotome, dermatome, and sclerotome. Epimere or somite consists of segments of mesoderm termed somites.
sperm (spermatozoan): Mature male gamete.
spermatid: Product of the meiotic division of a secondary spermatocyte.
spermatocyte: Stage in the maturation of the male gamete preceding the spermatid stage.
spermatogonium: Primordial sperm cell that gives rise to spermatocytes and sperm.
spinal cord: The portion of the central nervous system posterior to the brain.
spinal ganglia: Ganglia on the dorsal roots of spinal nerves.
splanchnic mesoderm: Hypomere mesoderm in close contact with endoderm.
splanchnopleure: The combination of splanchnic mesoderm and endoderm.
stomach: Enlarged section of the foregut posterior to the esophagus.
stomodeum: Ectodermal invagination that forms the mouth cavity.
stratum granulosa: Follicle cell–derived layer surrounding the inside of large ovarian follicles.
subcardinal anastomosis: Medial interconnection between left and right subcardinal veins.
subgerminal space: The space between the chick embryo blastoderm and the underlying yolk.

tail bud: Rudiment of the tail and posterior trunk of an embryo.

tail fin: Blade-like extension of the tail edge in amphibians.

telencephalon: Anterior portion of the forebrain.

theca externa: The outer layer of the amphibian ovary; the outer layer of mammalian Graafian follicles.

theca interna: A layer containing blood vessels, connective tissue, and endocrine glands, situated between the theca externa and the stratum granulosa of large ovarian follicles.

trachea: Tubular connection of larynx (or laryngotracheal groove) and lung bronchi.

tunica albuginea: Fibrous connective tissue covering the ovary and testis.

vegetal region (vegetal hemisphere):   Region of egg, opposite the animal region; yolk often accumulates in the vegetal region.

ventral aorta: Outlet of the embryonic heart that lies in the floor of the pharynx and conducts blood from the bulbus cordis (bulbus arteriosus, conus arteriosus) to the aortic arches.

ventral lip of blastopore: Lower or belly region adjoining the blastopore; last region to enter the blastopore during amphibian gastrulation.

ventricle (of heart):   Thick walled heart chamber that receives blood from the atrium. In the mammalian heart, the right ventricle delivers venous blood from the right atrium to the lungs via the pulmonary artery. The left ventricle delivers oxygenated blood (from the lungs and left atrium) to the body via the aorta.

ventricular septum: Muscular partition between the right and left ventricles.

wing bud: Wing rudiment.

yolk plug: The center of the circular blastopore in amphibian embryos, consisting of yolky endoderm cells.

yolk sac: Bag-like extraembryonic membrane extending from the midgut.

yolky endoderm: Large cells filled with yolk such as those in the floor of the amphibian midgut.

zona pellucida: A surface coat exterior to the plasma membrane of mammalian eggs.